Purposeful Evolution

Purposeful Evolution

RON DEMING

ISBN: 978-0-578-59493-4 (paperback)
ISBN: 978-0-578-59494-1 (ebook)

CONTENTS

CHAPTER ONE: Evolution

THE STAGGERING AMOUNT of biodiversity throughout Earth's history is attributable to biological evolution, which involves heritable genetic variation in populations over successive generations via mutations. Some ways that natural genetic mutations can emerge are through substitutions of a small number of nucleotides as well as DNA sequence frameshifts caused by deletions, insertions, or translocations. An accumulation of genetic mutations is evident through the fossil record because biodiversity increases from the oldest to the newest geological strata with the exception of mass extinctions. There is a lot of support for the theory of evolution from branches of science such as biochemistry, comparative anatomy, biogeography, comparative embryology, molecular biology, paleontology, biological anthropology, and geology. This book is not about determining whether evolution is true in comparison to an alternative such as creationism because this has already been explored ad nauseum. Instead, this book explores whether the genetic mutations needed for evolution are completely random or there is a degree of purposefulness involved.

The foundation of undirected evolution is complete randomness in regards to all genetic changes that have emerged throughout evolutionary history. Undirected evolution typically involves gradual changes because the emergence of genetic changes is unaffected by any influences in the environment such as environmental stressors that decrease the organism's survival chances. A problem facing the undirected emergence of beneficial mutations is randomness determining the aspect of an organism that is affected, which decreases the likelihood of appropriate genetic changes. This problem is compounded by the numerous outcomes that can affect each aspect of an organism because this further decreases the likelihood of an appropriate genetic change occurring.

The true extent of this problem surfaces when comparing the massive number of outcomes with the limited number that result in adaptability against stressors in a particular environment. The low probabilities involved with the emergence of a genetic change that is appropriate for conditions in an organism's environment is akin to a convenient coincidence.

Purposeful evolution happens when an environmental stressor triggers the emergence of genetic changes that increase an organism's survival potential against this particular threat. The balancing of ecosystems through purposeful genetic changes explains how so many well-adapted organisms exist despite the unique blend of hazards found in each environment. Threats to an organism's survivability that can trigger purposeful mutations include dangerous aspects of predators, characteristics of prey that help them against the organism, and factors associated with seasonal changes. Environmental stressors influencing the emergence of appropriate mutations explains how some species rapidly adapt to new environmental factors that negatively impact their survivability. The timely emergence of appropriate adaptations against a stressor found in an organism's environment is an indication that purposeful evolution is involved because this involves a high level of specificity. The degree of specificity linked to evolutionary change is important because the emergence of appropriate genetic changes in a timely manner indicates there is some degree of purposefulness involved.

Rapid evolutionary changes defy gradualism when appropriate genetic changes emerge for multiple species affected by the same environmental pressure. The eastern fence lizard (Sceloporus undulatus), lesser earless lizard (Holbrookia maculata), and little striped whiptail (Aspidoscelis inornatus) each evolved lighter skin that camouflages them in the white dunes of the White Sands National Monument. These are cases of convergent evolution because all three lizards evolved a similar evolutionary solution to an environmental stressor through the emergence of beneficial mutations. This is evident because the protein responsible for triggering the production of melanin has changed in different ways for all three species of lighter skinned lizards. The fact that a white-skinned variety exists for each of these species is remarkable because the

White Sands dunes have only been around for about 6,000 years. It would be a considerably unlikely coincidence for all three species to adapt to this specific environmental factor in such a short timeframe if it happened through undirected evolution.

Undirected randomness and purposefulness are not mutually exclusive propositions for the emergence of evolutionary mutations since there are many genetic changes that are most likely not influenced by environmental factors. Having an occasional mutation that is influenced by environmental factors does not detract from the numerous examples of neutral, harmful, and beneficial mutations that arise through undirected randomness. The false dichotomy of biodiversity being caused by undirected randomness or creationism is unfortunate because the idea of purposeful evolution does not fit into the narrative of either side. This false dichotomy will hopefully be shattered by evidence of a third option because purposeful evolution can easily coexist with undirected evolution. A proposition excluding the possibility of purposeful evolutionary mutations has a fatal flaw because it only takes a single instance of contradicting evidence to demolish the foundation of this argument. The burden of proof for supplying this evidence will hopefully be met through the ideas and examples expressed throughout these pages.

Natural selection is an essential part of evolution regardless of whether mutations are caused by a purposeful or undirected source. Natural selection happens when an organism's combination of physical and behavioral characteristics affects its survivability in a positive or negative way due to factors in its environment. Natural selection ties into evolution because new characteristics from mutations have the potential for increasing or decreasing an organism's survivability in its environment. Natural selection helps beneficial mutations spread throughout a population while limiting the reach of detrimental changes because a mutation's ability to spread is determined by its impact on survivability. Spreading a beneficial mutation is accomplished by organisms having more opportunities to reproduce over a longer lifespan which results in more members of the next generation having this beneficial mutation. The universality of beneficial changes increasing survivability and detrimental changes decreasing survivability means natural selection can operate no matter where a mutation occurs, which is important because every environment is different.

An example of natural selection is a genetic change in mosquitos that increased their survivability due to forming a resistance against a common insecticide known as DDT. The use of DDT was an excellent solution for killing mosquitos until a mutation in their GSTe2 gene allowed them to break down DDT into non-toxic substances. The development of this mutation increased survivability, so it gave those mosquitos more opportunities to reproduce and pass the beneficial genetic change to offspring. Interestingly, mosquitos with the mutation providing resistance to DDT are found in areas of the world that use DDT while it is not found in areas that don't. The genetic change was specifically helpful against a new factor in the mosquitos' environment, which indicates the environmental factor might have influenced genetic change. Further proof of purposefulness in this example is that the change was not gradual, given that it happened in an observable period rather than taking hundreds to millions of years.

An important distinction to make

is natural selection's inability to test a mutation before the genetic change has emerged through undirected or purposeful means. The emergence of a genetic change is critical to many instances of organisms adapting to a specific threat in their environment. This means the adaptability seen throughout nature is only possible when the proper mutations happen in the first place, so natural selection can subsequently determine the change's impact on survivability. If mutations account for how so many organisms have become well adapted to their environments, then the likelihood of these genetic changes emerging becomes pertinent. The likelihood of a mutation emerging from purposeful or undirected evolution can be difficult to determine because the context is usually unclear. It would help to know how long a threat was present before an appropriate mutation emerged. Both forms of evolution can potentially cause amazing instances of adaptability, but purposeful changes are typically rapid in comparison to undirected changes.

Convergent evolution is a strong indicator of purposeful evolution considering how appropriate the new mutations are for the various populations or species in relation to the commonality of environmental stressors. Convergent evolution is immeasurably important because it illuminates just how coincidental the independently emergent mutations are in comparison to the overall possibilities for mutations when undirected evolution is involved. Coincidences can only go so far before they are beyond believability. A form of convergent evolution dealing with populations of the same species is particularly informative in terms of purposeful evolution because it reveals that novel genetic changes are taking place. If two populations of the same species overcome a similar environmental stressor through different genetic changes, then the genetic change is novel rather than being preprogrammed into this species' genes. This indicates there are multiple genetic changes that can help overcome the same environmental stressor while also pointing to the influence of environmental factors for the emergence of appropriate mutations.

High altitudes found at the Qinghai–Tibetan Plateau, Andean Altiplano, and Ethiopian Highlands are associated with convergent evolution since populations at these locations experience the environmental stressor known as hypoxia. Short-term problems known to occur at lower oxygen levels include nausea, vomiting, impaired thinking, light headedness, muscle fatigue, blurred vision, and an accumulation of fluids in the lungs and brain. Long-term exposure to lower levels of oxygen can cause problems such as pulmonary hypertension, preeclampsia, and other pregnancy complications from decreased blood flow to the placenta. People can acclimate to these low oxygen levels when their body responds to hypoxia through their kidneys' release of the hormone erythropoietin, which increases the production of red blood cells. Some convergent mutations related to overcoming the problems of low oxygen levels seem to originate from adaptive introgression, which involves gaining adaptations through interbreeding with other species or subspecies. Adaptive introgression might involve purposeful evolution since these other species also experienced low oxygen levels.

People and animals at high altitudes show signs of convergent evolution when considering their adaptations to hypoxic environments involving a variety of genetic changes. Tibetans have adapted to low oxygen levels through changes to genes affecting the HIF pathway while Andeans have adapted through mutations impacting their cardiovascular system. Mutations to the EPAS1 gene have helped populations function in hypoxic environments including Tibetan humans, Tibetan wolves, Tibetan mastiffs, Andean horses, Tibetan hot-spring snakes, High Himalayan frogs, and snow leopards. Ethiopians and their cattle have developed adaptations to hypoxic stressors in a way that differs from Tibetans and Andeans, which involves genes associated with the HIF pathway such as VAV3. Similar adaptations to hypoxic conditions for organisms living in proximity to each other is evident again with the yak and Tibetan antelope through convergent evolution of the gene SOCS4. Appropriate adaptations can emerge rapidly as evidenced by Andean chickens developing mutations within a 500-year timeframe as measured by initial European contact.

Environmentally induced genetic changes that the idea of purposeful evolution proposes are not unheard of because a phenomenon known as phenotypic plasticity accomplishes this feat on a regular basis. Phenotypic plasticity involves variance in morphological and/or physiological characteristics of an organism through conditional changes to the expression of an organism's gene(s) as determined by environmental stimuli. Phenotypic plasticity helps organisms adapt by providing a way for some species to revert to morphologies and/or physiologies that were useful to the organism's ancestors facing similar conditions. Triggers for phenotypic changes can include environmental factors such as changes in temperature, climate, sustenance that is available, and nutritional levels in their diet. Plants can even experience phenotypic plasticity in the shape, size, and thickness of their leaves based on the exact levels of light in their environment. The ability to trigger latent morphologies and/or physiologies is important because it shows that appropriate genetic changes can happen based on specific circumstances in the organism's environment.

An organism that exhibits phenotypic plasticity through multiple instances of environmentally determined phenotypic outcomes is the pea aphid (*Acyrthosiphon pisum*). One outcome that is influenced by environmental factors is the determination of when generations of parthenogenetic females should be interrupted for the single generation each year that includes sexual individuals. There is only one generation each year that has males and oviparous females, which can subsequently create the overwintering generation that survives through winter due to diapausing eggs. The timing of the generation with two genders is triggered by the lengthening of nights during autumn and is crucial in protecting pea aphids from winter conditions and starvation. Pea aphids also have winged and wingless morphs that develop based on environmental factors, which in this case is triggered by the sustenance being provided by the host plant. Diminishing availability and quality of nutrients from the host plant influences the development of generations with wings that can easily travel to new plants.

A scientifically accepted phenomenon that is strikingly similar to phenotypic plasticity due to environmental factors impacting organisms at the genetic level is epigenetics. Epigenetics involves functionally relevant changes to phenotypic gene expression through regulation of the genetic code rather than alterations to the DNA sequence of the organism. Changes through genetic regulation are beneficial compared to altering the sequences of genes because adaptations to environmental stressors can occur within the organism's lifetime instead of in a future generation. The physiological phenotypic traits that emerge from epigenetic changes sometimes result in appropriate adaptations to stressors in the organism's environment that are having a negative impact on its survivability. An appropriate genetic response to a specific stressor in an organism's environment is an indication that a purposeful process is regulating genes due to a distinct lack of randomness. This theme of changes at the genetic level not involving complete randomness is unsurprisingly linked to a high level of specificity in external stimuli.

An aspect of environmentally influenced epigenetic changes that ties it closer to what is expected in evolutionary processes is that some of these alterations are heritable. Transgenerational epigenetic inheritance allows for an offspring's survivability to be impacted through heritable phenotype changes in the form of epigenetic tags that are based on current environmental factors. This can help improve the survivability of offspring because they are already adapted to current conditions without needing the same adaptations to emerge each generation. Transgenerational epigenetic inheritance shows that a genetic sequence does not always need to change in order to transmit environmentally influenced changes to offspring. However, these phenotypes can transition from flexible epigenetic tags to a longer lasting characteristic when it is encoded into an organism's genome through a process known as genetic assimilation. Genetic assimilation makes sense for purposeful evolution because the flexibility of epigenetic tags is unnecessary if the environmental factor that triggered the epigenetic change persists for an extended period.

There is criticism for epigenetic inheritance because the changes are not typically permanent for all future generations, so it does not adequately account for evolutionary changes. The problem with this criticism is that epigenetic inheritance should seldomly become permanent because environmental conditions can continue to change, so a lack of flexibility could negatively impact survivability. Epigenetic changes shouldn't become permanent if they are only beneficial to the organism in specific scenarios such as starvation or other situational conditions. Instead, epigenetic changes should react appropriately to changing conditions in the environment which makes the permanency associated with evolution counterproductive. Inherited epigenetic changes can result in maladaptive and detrimental outcomes which shows that these inherited changes aren't always beneficial. Detrimental outcomes can also happen when epigenetic changes are useful for the parent generation but are harmful for the new generation because environmental factors changed again. This shows the importance of natural selection acting upon changes and the nonrandom gradualism of genetic assimilation.

Inducible defenses are another indicator that genetic changes can occur based on environmental cues because these defenses involve genes being turned on or off based on the detection of threats. These conditional defenses are a form of phenotypic plasticity that allows organisms to conserve resources until damage occurs or a feature of a common threat is detected. Inducible defenses are only worth the expenditure of resources when a threat is nearby, so the production of appropriate countermeasures needs a regulatory mechanism that can accurately recognize threats. Some inducible defenses protect against a myriad of threats while others are specialized against a specific threat in the organism's environment. Inducible defenses are especially helpful for plants because their movement is limited, so they need another method for deterring attacks or indirectly dealing with these threats. One such method that some plants can use to indirectly deal with threats is producing a gaseous chemical that attracts predators of the pests attacking the plant.

The Lima bean (*Phaseolus lunatus*) is a great example for inducible defenses because they have a direct defense against a myriad of predators using the release of hydrogen cyanide when their cells detect damage. Lima beans have also developed an inducible defense against a specific threat in its environment known as the two-spotted spider mite (*Tetranychus urticae*). Being attacked by two-spotted spider mites causes the lima bean plant to produce Jasmonic acid, which subsequently induces the production of extrafloral nectar that attracts a carnivorous mite (*Phytoseiulus persimilis*). Lima beans also begin production of their extrafloral nectar when triggered by volatile organic compounds given off by lima bean plants that have been attacked by two-spotted spider mites. *P persimilis* have a mutualistic relationship with lima bean plants because the extrafloral nectar is an alternative food source for these mites. The lima plant only provides nectar for these carnivorous mites when two-spotted spider mites are detected, which the carnivorous mites also eat.

The reactive genetic changes associated with purposeful evolution can occur based on many factors within an environment including organisms that negatively affect survivability. These antagonistic relationships can lead to competitive coevolution wherein selective pressures from specific traits of two or more species affect each other's evolution. These evolutionary changes based on genetic responses to specific traits of another species can result in a reciprocal cycle of changes known as an evolutionary arms race. Having a reciprocal cycle leads to balance because each organism experiences changes to traits that result in advantages in the relationship until the other organism develops its reactive genetic change. Some organisms have maintained this cycle of alternating their advantage over each other by appropriately reacting to specific evolutionary changes over a span of many years. Multiple occurrences of coevolution over vast lengths of time show it is not a genetic fluke and that there are actual instances of evolution resulting from influences in the organism's environment.

An example of an evolutionary arms race that involves two species is the increased toxicity of poison used by rough-skinned newts (*Taricha granulosa*), which has been counteracted with newly increased resistance from garter snakes (*Thamnophis sirtalis*). This evolutionary arms race has resulted in the newt's poison, tetradotoxin, becoming one of the most potent neurotoxins. This neurotoxin is deadly because it blocks electrical signals between cells by binding to sodium channels in nerve and muscle membranes. The newt's poison effectively kills most organisms because it paralyzes muscles in the victim's diaphragm and heart. The other side of this evolutionary arms race involves the garter snake evolving effective countermeasures to this specific poison. The garter snake's countermeasure involves specialized modifications to their sodium channels, which makes it increasingly more difficult for the poison to affect them. Garter snakes increasing their resistance to the poison used by rough-skinned newts while this poison becomes increasingly potent shows a compelling connection between their genetic changes.

The Alcon blue butterfly (*Phengaris alcon*) is in an evolutionary arms race to maintain their parasitic relationship with various species of ants from the genus *Myrmica*. The cuckoo strategy's success heavily relies on mimicking the exact blend of hydrocarbons their host ants use to recognize each other. Secreting the correct blend of chemicals results in host ants bringing the caterpillar back to the brood chambers of their nest. An arms race occurs because larvae of these host ants secrete a slightly changed chemical blend that thwarts the caterpillars' mimicry. This is expected in terms of purposeful evolution because there is an environmental pressure for the host ants to adjust their chemical signature to stop Alcon caterpillars from stealing resources. The arms race continues when the Alcon caterpillars have an appropriate shift in their chemical mimicry and the host ants get further adjustments to their chemical blend. Butterflies escape the nest after emerging from their chrysalis thanks to loosely attached scales.

The contrast between undirected and purposeful mutations is prominent in cases of coevolution because there is a high level of specificity involved. It is important to note that natural selection plays a major role in the balance that emerges when species interact with each other during the process of coevolution. However, the question of whether specificity is involved does not pertain to natural selection because natural selection's role can't occur until there is a mutation for the environment to test. Mutations emerging based on specific traits of other organisms that can affect survivability and live within the proper proximity is what happens during coevolution. This is problematic because the probability of a mutation occurring that will alter the interactions with a specific trait of an organism within the proper proximity is extremely low. Therefore, organisms having a cycle involving an incrementally changed trait followed by reactive changes by the other organism is outside the realm of believability for undirected evolution.

The same emphasis on the importance of specific mutations emerging in response to pressures applies to the Rocky Mountain lodgepole pine (*Pinus contorta latifolia*) and two animals that eat its seeds. A threat to the Rocky Mountain lodgepole pine's seeds that has apparently influenced genetic changes due to direct interactions is the red squirrel (*Tamiasciurus hudsonicus*). The lodgepole reacted appropriately to the presence of red squirrels by making their seeds harder to access through changes such as heavier cones with thicker bases that contain fewer seeds. These genetic changes were effective against red squirrels because it became challenging for the squirrels to access the seeds inside the wider cones and more difficult to relocate the cones. The fewer seeds in these cones might be attributed to a shift in resources from creating more seeds to making access to these seeds more difficult. However, there is no evidence that the red squirrel has developed any mutations to counteract the lodgepole's changes.

A seed predator that has a cycle of coevolution with the Rocky Mountain lodgepole pine is the Cassia crossbill (*Loxia sinesciuris*). This species of crossbill has a specialized beak for eating seeds from lodgepole cones and consequently this is their only source of sustenance. The emergence of a species of crossbill with a specialized beak happened in areas without red squirrels, which influenced their scientific name because "sinesciuris" means "without squirrel". Rocky Mountain lodgepole pines that are in areas without the threat of red squirrels have appropriate changes such as cones with thicker scales at the tip. This is an appropriate change because the tip of the cone is where crossbills prefer to start prying the cone open to extract the seeds within. The Cassia crossbill has evolved in response to these thicker scales by developing a deeper and thicker bill that effectively counteracts the changes to the lodgepole's cones. This coevolutionary arms race shows that reciprocal adaptations can happen.

The low probabilities associated with mutations involved with antagonistic relationship are marginalized by the extremely low odds of genetic changes emerging that make some instances of cooperative coevolution possible. Cooperative coevolution involves separate species developing genetic changes to improve the survivability of one or more organisms in their mutualistic relationship. An extreme degree of interdependence can happen in these cooperative relationships wherein one or more members rely on specific characteristics of another organism in the relationship to survive. This is known as an obligate mutualistic relationship and likely involves purposeful evolution because organisms in the same environment are seemingly influencing each other's genetic changes. These interdependent relationships are also reminiscent of outcomes caused by purposeful evolution because they involve the increased survivability of one or more organisms in the same environment. Obligate mutualistic relationships are unlikely to happen if all genetic changes are the result of undirected evolution's complete randomness because these evolutionary outcomes involve a high degree of specificity.

Hummingbirds and various ornithophilous flowers are examples of cooperative coevolution since both sides have evolved specific characteristics that benefit their mutualistic relationships. Hummingbirds benefit from this relationship because the ornithophilous flowers they visit provide a sugar-rich nectar that is perfectly suited to the birds' diet. The flowers that accommodate the diets of hummingbirds also have the correct means of attracting these birds by using bright colors such as reds, yellows, purples, and oranges. Also, the flowers that provide the correct sustenance for hummingbirds usually lack a fragrance because these birds use eyesight to find food due to having little or no sense of smell. These flowers also accommodate hummingbirds by having flower blossoms situated for easy access and a structure that correctly matches the shape and length of the bird's bill. The hummingbird does not derive all the benefit in this relationship since the structure of the flower also causes the bird to assist with pollination after visiting multiple flowers.

An environmental factor that might trigger the process of purposeful evolution is the existence of nearby environments that currently have underutilized sources of food. Genetic changes based on underused resources involve mutations reacting to environmental factors with such a high level of specificity that the involvement of undirected evolution is extremely unlikely. This high level of specificity needed for the emergence of correct digestive systems would be a more believable coincidence if it was not so widespread. Genetic changes that allow an organism to digest an underutilized food source are an example of how purposeful evolution can result in balance throughout various ecosystems.

An organism that can't be digested by anything within a close enough proximity could lead to an imbalance because this particular population control would be lacking. Population controls exist in many different forms and each lends itself towards a balanced system because population controls complement other effective checks and balances within the ecosystem such as specialized features.

Genetic changes happening in reaction to the availability of new sustenance potentially explains how a strain of Flavobacterium developed the ability to digest byproducts leftover from the process of manufacturing nylon. Nylon is a modern synthetic polymer that was invented in 1935 and the discovery that a strain of Flavobacterium could degrade nylon oligomers happened in 1975. The protein that makes the breakdown of nylon oligomers possible is known as nylonase and has striking similarities to a carboxyesterase gene that can degrade nylon oligomers less optimally. Interestingly, there are only two mutations that are needed to achieve the enzyme nylonase from a pre-existing enzyme with similar capabilities. This pre-existing enzyme does not detract from the impressiveness of the emergence of nylonase because this optimization was accomplished with only two small mutations. Two minor genetic changes resulting in this strain of Flavobacterium's ability to effectively digest nylon byproducts shows a cause and effect relationship between new sustenance and purposeful genetic changes.

An example showing the emergence of genetic change based on available sustenance in incredibly rapid fashion is a bacterium that evolved to digest nylon byproducts in a lab setting. A strain of Pseudomonas aeruginosa developed the ability to break down the same nylon byproducts that the previously mentioned strain of Flavobacterium started digesting. This genetic feat was accomplished when the strain of Pseudomonas aeruginosa was placed in an environment devoid of any possible sustenance besides nylon oligomers. This was an incredible development because the strain of Pseudomonas aeruginosa did not have any enzymes capable of breaking down nylon oligomers before this test occurred. The emergence of specifically helpful genetic changes was different for the strain of Pseudomonas aeruginosa and Flavobacterium because the former accomplished this functionality with fewer nylonase genes. Furthermore, the emergence of specifically helpful genetic changes for the strain of Flavobacterium happened within a span of forty years while the strain of Pseudomonas aeruginosa changed in nine days.

Evolutionary changes are strikingly evident through the island rule that affects small populations when they encounter drastic changes in environmental factors due to being isolated on an island. These drastic changes in environmental factors through isolation have consistently been associated with appropriate genetic changes that increase the potential of survival for the affected population. One such genetic change is insular dwarfism wherein future generations of large animals have a noticeably reduced body size after becoming an island-dwelling population. A decrease in body size directly increases survivability because having a smaller body size means that each member within the isolated population requires less sustenance. The inverse is sometimes also seen through island gigantism, which involves a dramatic increase in body size of small animals when separated from a mainland population of the same species. It has been noted that changes in body size within a population of animals can happen rapidly by evolutionary standards when isolation on an island has occurred.

There are aspects of the island effect that provide better support for the idea of purposeful evolution rather than undirected randomness. It makes sense to imagine that undirected evolution could accomplish adaptations quickly in a smaller population because a beneficial change can spread to the entire population quickly. However, loss of diversity in a smaller genetic gene pool through the resultant population bottleneck is more likely to increase the time needed for appropriate genetic changes to emerge. There is a lack of gradualism in examples such as the Corsican red deer (*Cervus corsicanus*) that evolved a smaller body size in under six thousand years. Genetic changes emerging at this rapid pace makes sense in terms of purposeful evolution because the animals are reacting appropriately to environmental factors on their island. Limited resources on these islands might have triggered epigenetic changes that eventually resulted in genetic assimilation when these changes persisted through epigenetic inheritance because they remained beneficial in this environment.

It is important not to equivalate reactive evolutionary changes associated with purposeful evolution, which depend on environmental factors, with a process that results in perfect outcomes. Purposeful evolution results in adaptations that are specifically functional in the affected organism's environment and/or causes genetic changes that make a biological task possible rather than causing a perfect outcome. Natural selection can't exist in a system based on perfect outcomes because this eliminates detrimental genetic changes from the equation, which would also make adaptations based on environmental factors impossible. This is because adaptability for a characteristic of an organism can't occur based on environmental factors if this characteristic is already perfectly suited to every aspect of their environment. Instances of evolution that involve adaptations specifically helpful in an organism's environment alongside the demonstrable existence of natural selection shows a distinct lack of optimized outcomes. Optimal outcomes are also impossible because not all organisms in an environment can be optimized at the same time.

It makes sense that evolutionary changes caused by purposeful evolution have a tendency towards a balanced ecosystem rather than a utopia in which all organisms have achieved an optimal outcome. This is due to the fact that competitive relationships exist in nature including organisms that utilize the same sustenance and the dynamics formed by predators interacting with their various prey. Optimal outcomes are not possible in many ecosystems because new organisms are introduced, impactful events can occur such as climate change, and evolution periodically changes characteristics of predators and/or prey. An organism becoming better adapted against its predators and/or prey just shifts the balance until the new environmental pressure causes further reactive changes to occur. Ultimately, perfect outcomes are impossible in a system that requires some organisms to kill other organisms for sustenance. One of the best options for a system that requires predator and prey relationships is purposeful evolution because appropriate genetic changes can emerge to maintain balanced ecosystems.

CHAPTER TWO: Speciation

SPECIATION IS THE emergence of a new species that has reproductive barriers that stop copulation with another species from producing offspring that belong to either species. This is an incredible biological feat because it requires that the new species can't even reproduce with members of the species they directly branched from. This means a new generation is born that can't reproduce with members of the previous generation, so multiple members of this new species must exist for a successful speciation event. Differences in behaviors such as mating rituals don't result in the formation of a new species because members of both populations could still reproduce if proximity and/or preference changes. For example, a human not wanting to interbreed with another human because of their race, culture, political affiliation, or choice of hobbies does not make them members of different species. It takes more than a separation of populations for successful speciation events to occur because genetic changes are also involved.

The necessity of a genetic component for successful speciation events shows the importance of determining whether speciation is possible through undirected happenstance alone or if a degree of purposefulness is involved. Some aspects associated with speciation bring into question whether a purely undirected, gradual process could overcome significant challenges that potentially impede speciation events. On the other hand, it is also important to highlight the cause and effect relationships that might be influencing speciation events. The involvement of purposefulness would help explain the balance found in ecosystems despite catastrophic changes such as mass extinctions. This means the emergence of new species might be triggered by environmental factors such as underutilized resources and missing roles within an ecosystem. For instance, speciation could result in effective predators emerging to help reduce the chance of overpopulation. The balance in ecosystems coupled with the following problems facing new species point to the likelihood of a degree of purposefulness being involved in some speciation events

The impact of evolutionary changes within a population is limited by how many members are affected by each genetic change that emerges. This feature stops beneficial characteristics from affecting an entire population, but is crucial because a single deleterious genetic change is stopped from immediately affecting enough members to eliminate a population. Instead, the spread of genetic changes is determined through impacts on survivability through natural selection, which protects against widespread harmful effects from a single genetic change. The problems associated with gradualism when multiple genetic changes are needed for a successful speciation event are exasperated by the limited number of members affected by each genetic change. A limited number of members of the parent species must be affected by multiple genetic changes so the combination of changes results in a new species rather than causing adaptations. This concentration of relevant genetic changes branching from a limited number of members within the parent species demonstrates specificity rather than undirected randomness.

This problem facing undirected evolution concerning the limited number of members within a population affected by each genetic change could be alleviated by having a smaller population. This potentially helps because having a smaller population makes it easier to have a concentration of appropriate genetic changes needed for a successful speciation event within the correct members. However, a smaller population has its own problem in the form of limited members that can be affected by the genetic changes required for a successful speciation event. Having less opportunities for mutations to occur means it is less likely for the exact combination of genetic changes necessary for a speciation event to emerge. A change in the rate of mutations based on population size indicates that environmental factors are influencing the emergence of genetic changes. If environmental factors can't affect the rate of mutations, then it decreases the likelihood of appropriate genetic changes developing within the smaller populations usually associated with speciation events.

Purposeful evolution makes sense for speciation events to occur within populations that have been transplanted to new areas or have otherwise been impacted by significant changes within their environment. These significant changes can create an imbalance in the old and/or new environment which triggers speciation events to adequately refine a balance within this ecosystem. For example, a species of flower that relies on a specific pollinator might be transplanted to an area that lacks this symbiotic partner. This species of flower would not have the chance to propagate further generations in the new environment unless significant evolutionary changes occurred within the lifespan of the transplanted population. It is obvious that these significant changes sometimes occur because flowers in each environment throughout the world have pollinators that appropriately match their anatomy. The proliferation of flowers with different pollination needs might have happened when pollinators and/or flowers developed appropriate features or a new species emerged to better fulfill needs in their environment.

An indicator of some degree of purposefulness existing outside genetic changes is parents taking care of their offspring when the first generation occurred for species with dependent offspring. Parents of dependent offspring within this first generation of a species wouldn't have known of their offspring's reliance on others for obtaining sustenance if at least some purposefulness didn't exist. This problem is unavoidable because there had to be a shift from independent to dependent offspring at some point in the history of speciation events. This crucial aspect of parents caring for their dependent offspring requires further purposeful intervention when the needs of the offspring differ from those of its parents. Purposeful intervention is needed because the needs of the offspring must be fulfilled by resources within an appropriate proximity and the parents must figure out what their offspring's needs are. If the parents didn't figure out the specific needs of the genetically changed generation, then the speciation event would have failed.

There is a distinct lack of motivation for parents to take care of their dependent offspring unless a degree of purposefulness is involved. The dependent offspring would likely be competitors for the same resources that the parents use, so it makes no sense for the parents to share these resources willingly. There was a lack of precedent for providing help to another organism at the expense of the provider when the first generation of dependent offspring emerged. A shift in instincts would need to occur from focusing on the organism's own survival needs to also providing for the needs of its offspring. There might have been a preexisting instinct for wanting to help the continuation of the species through providing help, but this makes no sense when each organism was already self-sufficient. Wanting the continuation of the species seems to be a purposeful instinct because it makes no sense for this instinct to exist before the first dependent offspring emerged.

Having purposeful instincts would be irrelevant for successful speciation events if dietary needs of genetically altered offspring were not properly met. It would not matter if the parents recognized that their offspring had different dietary needs if the required sustenance did not exist or was not within proper proximity. Changing dietary needs are a critical problem facing undirected evolution because of the specificity required for each organism to emerge near an environment with appropriate sustenance. An organism does not have enough time to sample all of the available sustenance to find the ones it can digest properly given their limited mobility and lifespan. Having the exact digestive changes to match available sustenance is an incredible feat that has been accomplished many times given the biodiversity that has existed on Earth. This is a major coincidence considering the number of digestive systems appropriate for currently available sustenance compared to the number of digestive systems that wouldn't match any currently available sustenance.

A species that illustrates the dietary problem facing speciation is the marine iguana (*Amblyrhynchus cristatus*) from the Galápagos Islands. The marine iguana's diet consists exclusively of underwater algae and seaweed, which is narrowed even further because they can eat green algae while lacking the ability to digest brown algae. They are well-adapted for efficiently swimming to their sustenance using a flattened tail and gathering their food by scraping algae off rocks with their razor-sharp teeth. It does not matter if this dietary change happened after this species already emerged because the dietary problem would be the same regardless of when the changes to the digestive system occur. Changes might have happened to their digestive system in a way that allowed marine iguanas to eat the same sustenance as their parents in addition to the algae and seaweed. This possibility does not detract from the specificity needed for marine iguanas to develop a digestive system that matches the exact sustenance available.

Italian wall lizards (*Podarcis sicula*) show how organisms can adapt rapidly and accordingly to environmental factors such as changes to available sustenance. Five adult pairs of Italian wall lizards were intentionally transplanted from their home island to a nearby island in 1971 and were not studied again for about 36 years. Genetic changes observed in this transplanted population included an increased bite force from their head becoming longer and wider. These lizards also developed a change to their gut structure through the development of a cecal valve which helps them digest vegetation more effectively. These specifically helpful genetic changes helped the lizards' transition from eating a diet of mostly insects to taking advantage of the abundant plant life on the new island.

This means the genetic changes might have been triggered by changes in the lizards' environment. The rapid changes made by these transplanted lizards help explain how each organism is well-adapted to the exact sustenance available in their environment.

Speciation events involving two sexes are reliant on coordinated mutations for their reproductive systems because genetic changes can't be passed to a new generation unless compatibility is maintained. The fact that this level of specificity has been achieved enough times for millions of species to emerge throughout Earth's history is incredible considering the extreme selection process involved. An organism with a mutated reproductive system must overcome typical environmental threats combined with the possibility of not having compatible mates. A substantial hurdle must be overcome wherein a new species becomes reproductively isolated from the parent species regardless of whether these mutations emerge through a one-step transition or gradual phases. Members of the new species must then learn to differentiate between compatible and incompatible mates for a speciation event to have a higher rate of success. The need for locating appropriate mating partners is part of the reason why higher rates of speciation events are attributed to relatively small and isolated populations.

Environmental factors can affect the gender of some organisms as evidenced by clownfish changing their reproductive functions based on how many female clownfish are nearby. Their ability to transition from male to female based on a group's lack of a female makes reproduction possible since all clownfish are born males. The process of transforming from a male to female is based on specific environmental conditions such as being the most dominant male when the female of the group dies. Having specific criteria for the transition from male to female is important because the random emergence of females would disrupt the hierarchy that exists within each school of clownfish. The new female of the group chooses a dominant male for a breeding partner, and this mate will likely become the next female when the need arises. This process makes evolutionary sense because the most dominant members of each group are the only members that contribute their genes to the next generation.

Purposeful evolution is evident through

temperature-dependent sex determination because reproductive functions are being affected by an environmental factor rather than having an outcome that is only determined genetically. The sex of green sea turtles (*Chelonia mydas*) is determined by the temperature of their nesting environment after fertilization has occurred. Males are more likely to hatch when temperatures are lower while more females hatch when the nesting environment is warmer during the period of time when sex determination happens. Recent increases in global temperatures have caused a disproportionate ratio of females to males because their sex is determined by environmental conditions. Their prior survival through approximately 100 million years of temperature fluctuations might have been accomplished by changing when eggs are buried in the sand and/or the location of their nests. Having a determination of sexes through external stimuli such as temperature is a strong indicator that purposeful evolution exists because undirected evolution likely wouldn't result in environmentally derived genetic outcomes.

The common slipper snail (*Crepidula fornicata*) demonstrates that sex determination can be affected by external stimuli, which goes against a system based on genetic changes that are completely undirected. Common slipper snails become male when they mature and can become females based on environmental circumstances. The transition of common slipper snails is determined by the snails' positioning in relation to nearby peers and the sex of the snails it is stacked with. These snails form mounds by stacking on top of each other and the bottom individual is invariably going to become a female because of its location in the mound. The next snail on the stack will have the reproductive role of a male until another snail joins the stack, which causes the lower male to transition into a female. Sex determination based on position in the mound makes evolutionary sense because reproduction will continue as long as there are two members of the species within the proper proximity.

The occurrence of sex determination based on the snail's location in the mound shows that at least one aspect of the organism's genetic outcomes is directed by environmental factors. If one genetic outcome is directed by external stimuli, then it opens the possibility that other genetic aspects can be influenced. Having sex determined through environmental factors creates further deviations from a system based on purely undirected outcomes, which serves to further undermine the validity of undirected evolution existing alone. This problem for undirected evolution is derived from the onset of a system that involves environmental factors rather than undirected randomness because it requires a shift in how genetic outcomes occur. To better understand the dilemma for undirected evolution, imagine if flipping a coin had its outcome determined by temperature or location of the coin flipper rather than based on probability. This would eliminate the involvement of randomness when flipping a coin just like undirected evolution is undermined by environmental influences.

Proximity is a problem facing successful adaptations and speciation events and is interwoven into other problems such as dietary changes and the emergence of reproductive systems for a new species. Members of a new species must emerge within the correct proximity so they have a reasonable chance of interacting to make new generations possible. It does not matter if appropriate changes are made to a digestive system to match a specific food source if that sustenance is not close enough to utilize for nourishment. Purposeful evolution does not make sense in regards to a genetic change unless the mutation happens in response to an environmental stressor that has appropriate proximity to the evolving organism. It is not purposeful evolution if mutations are causing adaptations to a change in an environment that is thousands of miles away from the organism's habitat. Proper proximity for genetic changes that are appropriate for the current conditions is an indication that purposeful evolution has occurred.

Timing is another problem that is consistently interconnected with every other problem facing the emergence of appropriate mutations during periods of adaptive changes and speciation events. The emergence of a digestive system that matches an organism that lived in the proper area is irrelevant if the other species went extinct a million years earlier. It also does not matter if each reproductive system emerges for a successful speciation event if both sexes don't exist while each member is still in their reproductive stage. Timing is a bigger problem for undirected evolution than proximity because the coordinated emergence of reproductive systems that makes a new species possible brings the involvement of gradualism into question. Timing is a strong indicator of purposefulness when appropriate genetic changes occur after a new stressor is introduced or a population is transplanted to a new area. Timing is obviously not an insurmountable problem given how many species exist and how well-suited they are for their environment.

The problems of proximity and timing are acute when a species has an extremely short lifespan because it places limits on the time gradualism has to operate. The species of mayfly demonstrate this conundrum because the sexually mature stage of mayflies last for a maximum of two days. Extreme cases like the various species of mayfly reproducing for such a short time period help illuminate the problems facing speciation. A new species of mayfly needs to develop two compatible sexes while also needing to find a mate from the same new species in two days or less. The problem of timing is alleviated slightly because batches of mayflies reach their sexually mature stage at the same time. This still seems like an impossibility for gradualism given the inability to hereditarily pass on intermediate reproductive changes for further refinement. This might mean gradualism is not involved in the speciation events of mayflies since over 2,500 species exist despite the challenges involved.

Cospeciation happens when multiple species each undergo their own successful speciation event while resulting in compatible genetic outcomes for the species involved. Cospeciation requires that each species overcomes all problems associated with normal speciation while sometimes needing the coordination of multiple speciation events. This coordination is critical for obligate relationships wherein both members of the partnership rely on each other for survival because the speciation events must occur within a limited timeframe. Cospeciation is less problematic when only one partner is reliant on the relationship because they can continue interacting with the other partner's parent species until their own speciation event happens. Any examples of coordinated speciation events are extremely damaging to the believability of undirected evolution because it stretches the bounds of what is reasonably possible with undirected randomness. The low probabilities involved with undirected evolution overcoming the problems of speciation are exponentially worsened when considering the odds of coordinated speciation events when the species have an obligate relationship.

Cospeciation is noticeable when species have a close association such as the parasitic relationship of lice and the species of gopher they use as a host. Lice don't have many opportunities to switch hosts so genetic changes and transmissions are typically isolated to the lice that have a parasitic relationship with the same species of gopher. The emergence of a new species of gopher results in isolation of lice that use them as a host, which makes it easier to compare genetic changes of lice populations. A determination that cospeciation has occurred is easier with parasitic relationships wherein interactions between the species persist for generations because genetic changes can be reasonably linked to these close associations. Cospeciation is apparent when the phylogeny of lice and their host gopher share strong similarities because their respective parent species also have a host/

parasite relationship. Close associations between species might help cospeciation because frequent interactions provide an opportunity for species to develop coordinated changes.

A common trend found throughout the fossil record is organisms that show minimal evolutionary changes while new species appear suddenly as if their speciation events happened rapidly. This is known as punctuated equilibrium and focuses on the prevalent stasis found in the fossil record that can last for hundreds of millions of years for some species. Not all genetic changes are impactful enough to an organism's features, so it is expected that not all mutations are captured in the fossil record. However, some transitions that occur during a speciation event require multiple morphological changes that could be captured by the fossils forming during the speciation event. It is not as if morphological changes are not captured in the fossil record because these noticeable changes are one of the ways speciation events are recognized. A problem with capturing speciation events is how rarely organisms turn into fossils, which explains why a rapidly speciating species was unlikely to leave behind fossil evidence.

The existence of stasis in the fossil record and new species appearing abruptly are both strong indicators that undirected evolution is not the only cause of speciation. Prevailing stasis intermixed with bursts of speciation in the fossil record undermines the believability of purely undirected evolution because the rate of mutations should not fluctuate so dramatically. Fluctuation is evident because some organisms persist for millions of years without noticeable changes while species emerge so quickly that each individual morphological change is not captured in fossilized form. It is possible for the fossil record to not capture a speciation event involving gradualism given the particular conditions needed for a fossil to form. However, it is more likely for the fossil record to capture changes made by undirected evolution because the changes for a speciation event would require a substantial amount of time. The gradualism tied to undirected evolution is not what is indicated through stasis and abrupt speciation in the fossil record.

Punctuated equilibrium makes more sense in terms of purposeful evolution because stasis and abrupt appearances of species support the appropriate genetic reactions to environmental factors. The abrupt appearance of new species throughout the fossil record points to a process such as purposeful evolution that does not operate gradually. The sudden appearance of species in the fossil record makes sense because speciation could happen quickly in response to environmental factors such as available sustenance. Environmental factors that affect an organism's survival can remain stable for long periods of time if impactful events such as volcanic eruptions and extreme climate change don't occur. Species having a lack of pressure to trigger an adaptation response due to being well-adapted to their current environment can lead to extended periods of stasis for various species. Adaptations and speciation being determined by nearby environmental factors explains why some species look similar to their ancient fossil counterparts while new species can rapidly emerge with multiple morphological changes.

A new species that exploits a previously underutilized ecological niche is an indicator that purposeful evolution is involved with that speciation event. Underutilized resources should trigger an appropriate evolutionary response to ensure balance is maintained in each ecosystem. A tephritid fruit fly known as the *Lonicera* fly fits this description because its homoploid hybrid speciation event roughly coincided with the introduction of nonnative honeysuckle species from the genus *Lonicera*. The emergence of the *Lonicera* fly was likely not a gradual process given that these species of honeysuckle have only been growing in North America for 250 years. This speciation event probably would have gone unnoticed if each species of tephritid fruit fly was not known for their recognition and exploitation of a particular plant host. This homoploid hybrid speciation event created a fertile species with approximately a hundred generations of genetic divergence, which means this ecological niche hasn't been filled through the repeated interbreeding of *Rhagoletis zephyria* and *Rhagoletis mendax*.

An idea that encapsulates what purposeful evolution is capable of in regards to underutilized niches is adaptive radiation, which involves multiple rapid speciation events stemming from a single ancestor species. Speciation events associated with adaptive radiation are demonstrations of purposeful evolution in action because each new species is specialized to utilize the resources available in the new environment. Adaptive radiation's speciation events point to quick evolutionary responses to specific environmental factors. On the other hand, adaptive radiation should not be possible if evolution is only based on undirected randomness because this process shows a distinct lack of gradualism and randomness. Gradualism is defied by adaptive radiation's rapid emergence of multiple species because a gradual speciation event should take a significant amount of time considering how many genetic changes are needed. The viability of Undirected randomness being involved is undermined because each new species develops the exact morphological changes appropriate for available resources despite the high number of possible genetic changes.

The thirteen species of finches living on the Galápagos Islands demonstrate what adaptive radiation is capable of because they all branched from a single ancestor on islands formed by volcanoes. The oldest Galápagos Island is dated to around 3.3 million years old, which means this is the longest period of time these finch species had to branch from their ancestor. Each species of finch has a specialized beak that allows the various species to coexist on the islands with minimal overlap in the food sources they exploit. For example, the different ground finches have beaks that allow them to eat the available seeds with some overlap. Variation becomes important when food becomes scarcer because each species can survive on the small, medium, or large seeds their beak is specialized for. Darwin's finches represent some of the best examples against what undirected evolution is capable of since the species emerged rapidly while each developed a specialized beak matching available food sources.

The fast speciation events of adaptive radiation are contrasted by the short time it takes for species to disappear during mass extinctions. Mass extinctions are not problematic to the idea of purposeful evolution even though genetic changes don't emerge quickly enough to prevent the mass extinctions from occurring. The ability of purposeful evolution to help species to emerge and adapt based on environmental factors is insufficient for counteracting the multiple cataclysmic events that happen at a rapid rate. Purposeful evolution might have the ability to counteract one cataclysmic event, but impacts on survivability during events such as the Cretaceous–Paleogene (K–Pg) mass extinction had interconnected consequences. The intertwined consequences of the space rock impact in the Yucatán Peninsula, volcanic activity in India, and climate change caused the extinction of approximately 75% of species on Earth. Another contributing factor to the mass extinction was the resultant dust cloud from the space rock's impact blocking sunlight for up to a year.

The loss of sunlight for an extended period of time or other impactful environmental changes can cause a domino effect of extinctions as an increasing number of food sources disappear. The extinction of the only species that pollinates a plant will directly lead to the extinction of the plant unless adaptations are made, which are highly unlikely during mass extinctions. Insects and other animals that rely on the extinct plant for sustenance must make appropriate changes to their digestive systems to utilize a new food source or extinction might occur. The domino effect of extinctions worsens as more imbalance happens in an ecosystem because even species that utilize multiple sources of sustenance can eventually lose all sources of food. The interconnectedness of food webs helps create balanced ecosystems that make it possible for speciation events to fill specific roles. This same interdependence amongst species makes mass extinctions more devastating than they would be if species did not have specialized roles.

The worst mass extinction in Earth's history happened about 252 million years ago and wiped out approximately 96% of marine creatures and 70% of terrestrial vertebrate species. This is known as the end-Permian extinction and there is evidence that this extreme loss of biodiversity happened very rapidly according to geologic records. Such a drastic decrease in biodiversity should have happened quickly and had multiple causes if purposeful evolution couldn't help prevent extinctions by helping species adapt to the detrimental environmental changes. This is indeed the case in regards to multiple causes with one of the biggest contributors to the extinction event being massive volcanic eruptions in what is now Siberia, Russia. Ocean life was critically impacted from greenhouse gasses released by the volcanic activity because they increased acidity and decreased oxygen in the oceans. The problems of increased acidity and decreased oxygen in the oceans was compounded by a large population boom of anaerobic *Methanosarcina,* which released methane as waste.

Mass extinctions are an extreme problem for undirected evolution because they amplify problems such as speciation events needing the correct proximity to sustenance and needing an appropriate digestive system. A speciation event resulting in a digestive system that is appropriate for available sustenance despite the many possible genetic outcomes is difficult under normal conditions. The digestive problem is amplified to an absurd level when considering the number of appropriate digestive systems after a mass extinction has wiped out the majority of species on Earth. The problem of proximity is similarly affected because the likelihood of a new species emerging in the correct proximity of sustenance matching its digestive system is exceedingly reduced. The digestive and proximity problems are acutely pronounced for speciation events that occurred in the oceans after the end-Permian extinction. Mass extinctions increase the level of coincidence involved with undirected evolution successfully developing the exact genetic changes needed for a new species while also overcoming speciation's other problems.

On the flip side, biodiversity returning after each mass extinction is a strong indicator that purposeful evolution can still operate effectively when major stressors are affecting an ecosystem. The major stressors that purposeful evolution can alleviate are the ecological niches that are disrupted after a significant number of species have gone extinct. Adaptive radiation can rapidly replace missing roles in ecosystems when conditions have improved enough to accommodate an environment of checks and balances with normalized fluctuations. Opportunities for speciation events increase as biodiversity increases because there are more resources to exploit for a new species with an appropriate digestive system. This creates a ripple effect wherein the emergence of new species provides triggers for further species and specializations to emerge based on the dynamics found in a given ecosystem. This means mass extinctions are not problematic for purposeful evolution because balance can be reestablished over and over again through appropriate genetic changes and speciation events based on environmental factors.

CHAPTER THREE:
Plants and Animals

Some examples of plants and animals are more effective at demonstrating the potential involvement of purposeful evolution through the existence of adaptations that are effective against factors in the organism's environment. Determining whether mutations are caused by undirected randomness or purposefulness relies on information that is not always available, such as environmental conditions that existed when the genetic change emerged. This determination is easier when multiple mutations have emerged that are specifically appropriate in an environment despite the numerous genetic outcomes that could've occurred if evolution is purely undirected. Some appropriate mutations stretch the believability of genetic outcomes as pure coincidences. If a single genetic outcome is beyond what a purely random process could achieve, then it undermines the believability of undirected evolution as the only contributing factor for emerging mutations. Decide for yourself if the following examples involving specificity are likely to occur through purely undirected randomness or whether it is more likely influenced by a degree of purposefulness.

Koalas (*Phascolarctos cinereus*) are tree-dwelling marsupials that live in areas of Australia that have the correct eucalyptus trees for their diet. Koalas have several adaptations that allow them to survive on a diet of mostly eucalyptus leaves, which are poisonous and not very nutritious. They have sharp front incisors for removing leaves, specialized back teeth for grinding eucalyptus leaves, and a gap between the incisors and molars so their tongue can rotate the leaves. The low energy gained from the leaves is appropriately counteracted by a body temperature that is consistently lower than most mammals and a low metabolic rate that helps conserve energy. Koalas have a smaller brain in relation to their size in comparison to closely-related marsupials, which makes it possible to survive on eucalyptus leaves because their brain needs less energy. The koala's seven-foot cecum allows them to maximize the energy extracted from eucalyptus leaves while specialized bacteria in the koala's stomach detoxify toxins in the leaves.

Various species of anglerfish rely on parabiotic reproduction, which involves a male anglerfish fusing with a female anglerfish host after he bites into her skin. The male releases a specialized enzyme that dissolves his mouth and the female's body at the bite mark before they fuse together at the blood-vessel level. Female anglerfish provide nutrients for any males that are attached to them while males only provide sperm. The males of some species of anglerfish would not survive long enough to reproduce without parabiotic reproduction because they have jaws that are not effective for capturing prey. The survival dilemma is increased for species that have males that lack a properly working digestive system. This creates a conundrum for undirected evolution because the fusion process has no advantage in terms of natural selection if males can capture and digest prey. On the other hand, species of anglerfish would likely go extinct before this specialized process emerged if the males couldn't eat.

The blue mussel (*Mytilus edulis*) has an inducible defense wherein it thickens its shell when it detects the chemical cues of green crabs (*Carcinus maenas*). The chemical cues of this specific species of crab trigger an appropriate inducible defense because these crabs can normally crush the shells of blue mussels. However, this inducible defense is not helpful if a predator within the vicinity of the blue mussel does not trigger the thickening of their shell. Having an inducible defense that does not trigger when a threat is present was the dilemma facing blue mussels when Asian shore crabs (*Hemigrapsus sanguineus*) invaded part of their habitat. It took fifteen years or less to overcome this problem because it took this long before someone noticed blue mussels' shells thickening in response to chemical cues from Asian shore crabs. Rapid inclusion of new chemical cues for triggering an inducible defense shows that new threats might cause appropriate genetic changes in a timely manner.

Guppies (*Poecilia reticulata*) are small tropical fish found in freshwater that have been found to adapt in a decade or less to the presence or absence of predators. Research has been done by transplanting populations of guppies to see how they adjust to the new environment and how their presence affects the ecosystem. The presence or lack of predators has a noticeable impact on factors such as the rate at which guppies mature and the number and size of offspring produced by each female. Guppies in an environment that lacks predators for their adult population adjust to these conditions by maturing slower and females produce fewer, but larger, offspring. This is beneficial in this environment because larger juvenile guppies are less likely to be the victim of a predator. Guppies located in an environment with predators for their adult population mature faster and produce more numerous offspring that are smaller, which helps because there might be less opportunities to reproduce.

Daphnia magna is a water flea that can live in a variety of fresh and brackish water environments. These water fleas alternate between the production of offspring through asexual reproduction when the environment has favorable conditions and sexual reproduction through fertilized haploid eggs when certain environmental changes occur. Conditions that trigger the production of male offspring include the lengths of daylight and darkness in fall that indicate winter is approaching, increased population density, and a shortage of sustenance. The production of males is important when adverse conditions occur because the fertilized eggs enter diapause until conditions improve and these eggs are protected by an extra shell layer. There must be at least one trigger to cause a switch from reproducing asexually to sexually or there would be no reason for females to alternate between modes of reproduction. Environmental conditions must also affect when eggs in diapause hatch so they don't hatch before conditions become favorable again or potentially never hatch.

The Arctic fox (*Vulpes lagopus*) is a small mammal that is well-adapted to the harsh conditions of the Arctic tundra. The Arctic fox is camouflaged all year because they blend in with the tundra rocks during the summer with brown or gray fur and transition to white fur during the winter. The two layers of white fur that Arctic foxes have during the winter provide more insulation from the cold than the fur of any other animal in the world. Avoiding the loss of body heat is helped by a compact body shape combined with shorter legs and ears because there is less surface area exposed to the Arctic cold. Another way that body temperature is regulated is maintaining different temperatures for their core body and the part of their paw that is in direct contact with the ground. This makes sure the pads of their paws stay above the tissue freezing point while minimizing their loss of body heat.

The three-toed skink (*Saiphos equalis*) is a lizard native to eastern Australia with multiple reproductive modes that differ based on their environment. The three-toed skinks that live in the warm coastal lowlands lay eggs and members living in the colder northern highlands tend to give birth to live young. Determining reproductive modes based on temperature is an appropriate adaptation because members lay eggs in areas that provide steady warmth for embryotic growth and have live births in colder areas. The option of having live births instead of laying eggs can also be beneficial when predators are nearby because eggs are vulnerable to attack. Embryos held in their mother until birth miss out on calcium from an egg's shell but this is fixed by mothers secreting calcium from their uterus while embryos grow. The three-toed skink is an example of populations of the same species having differences based on environmental conditions that might not be discernable through studying their fossil records.

Male oceanic field crickets (*Teleogryllus oceanicus*) have specialized wings to create the distinctive chirping that is used to attract mates. The parasitoid flies (*Ormia ochracea*) that arrived on the Hawaiian island of Kauai in 1991 can accurately pinpoint the location of chirping crickets and follow the sound to its source. The flies home in on the location of the chirping sound and use this male cricket as a host for their larvae. It took under twenty generations for oceanic field crickets to adapt to this problem through a genetic mutation that changed the morphology of male wings enough to render them silent. Oceanic field crickets on the neighboring island of Oahu developed a similar "flatwing" trait that effectively silenced males through a different mutation when the parasitoid flies reached their island. It is incredible that a similar genetic change emerged for two separated populations of oceanic field crickets in a timely manner after each population encountered the same environmental stressor.

The numerous cichlid species are brightly colored fishes that provide a great example of specialized features helping fill niches through adaptive radiation. Hundreds of cichlid fishes fill many roles in large lakes found in East Africa because of their morphological variations and because they're secondary freshwater fish in environments with higher salinity levels. Cichlid fish have a second jaw known as the pharyngeal jaw apparatus, which is usually morphologically different for each species and appropriately matches the food located in that fish's habitat. Some cichlid fishes have pharyngeal jaws with phenotypic plasticity, which results in their jaw morphology being determined by available sustenance. Cichlid fishes eat foods that match their specialized jaw, and sustenance for the various species encompasses a wide variety including plankton, insects, algae, snails, and body parts of other fish. Specializations found in jaws of cichlid fishes have the benefit of reducing competition amongst species in an environment because there is reduced overlap in their preferred sustenance.

Carnivorous plants have adapted to

environments with low nutrients by trapping and digesting insects, frogs, and other small animals. The ability to digest animals is an appropriate change in environments with insufficient nutrients because it allows carnivorous plants to take advantage of underutilized niches. This is evident because carnivorous plants are found in bogs, fens, and other areas where most plants would struggle to survive because the environment lacks sufficient nutrients. To transition from obtaining nutrients from the soil to digesting animals as a food source requires genetic changes that help the plant trap the animals while they are being digested. Carnivorous plants also need mutations for extracting nitrogen from animals when they touch the plant because the trap itself does not overcome the lack of nutrients in the environment. It isn't necessary for a proper trap and digestive system to emerge simultaneously, but carnivorous plants would struggle to obtain nitrogen if their prey doesn't stick around to be digested.

Pitcher plants are a type of carnivorous plant that have multiple features that help them gain nutrients from small animals such as insects. The most noticeable feature is the specialized leaf that grows into a hollowed-out structure with a rim at the top and pool of liquid at the bottom. An important aspect of the pitcher plant's system for getting nutrients is the production of nectar that entices its prey to approach the pitfall trap. The animals might slide into the pool of liquid at the bottom because the upper portion of the pitfall trap directly below the rim has a wet or waxy surface. There are also hairs growing at a downward angle within the specialized leaf helping prevent animals from escaping the pitcher plant's trap after they have slid into the liquid pool. The pool of nectar contains enzymes, secreted by the leaf or bacteria from an external source, that digest victims to obtain nutrients such as nitrogen.

Butterwort is a common name shared by eighty species in the same genus of carnivorous plants that need moist or wet environments and thrive in shady or partly sunny areas. Different species of butterworts can survive in a range of harsh conditions including high alkaline soils, soils of nearly pure gypsum, acidic peat swamps, and vertical rock walls. Some species can even survive bone-dry conditions by abandoning their resource-draining carnivorous leaves and replacing them with succulent leaves until conditions improve. Butterworts trap their prey by producing tiny droplets of mucilaginous secretions that give the appearance of moisture on the leaves that can lure thirsty insects close enough for physical contact. The insect's attempts to escape the sticky substance just cause the secretion of more mucilage from the plant's special reservoir cells. The sticky droplets have small amounts of digestive enzymes and the initial intake of nitrogen from an insect triggers the secretion of more enzymes that finish the digestion process.

Successful reproductive plants have pollination systems with features that are appropriate for pollinators in the correct proximity or conditions in their environment. Some plants have specialized features that effectively attract the correct species of pollinators that complement their particular pollination system. This is a helpful evolutionary outcome because pollination systems require pollen from the same species of flowers, which increases the benefit of pollinators visiting a lower number of flower species. Pollination specializations are sometimes accomplished through having colors, scents, and other cues that increase the reliability of pollinators visiting multiple flowers within the same species so reproduction can occur consistently. The likelihood of achieving a specialized pollination system is hindered by the number of pollinators in the correct vicinity that already pollinate enough flowers to fulfill their own needs. The problem of developing an appropriate attracting factor is exacerbated by needing to attract a pollinator that has physical characteristics that match the features found in the flower's pollination system.

Hydnora africana is a parasitic plant that is missing conventional features of plants such as leaves and chlorophyll and mostly grows underground with the exception of its above ground flower. *H. africana* gets nutrients exclusively through an obligate parasitic relationship with host plants from the genus Euphorbia, which means it can't survive without stealing nutrients from these specific host plants. They have a specialized enzyme to dissolve some of the host plant's roots so it can attach and extract nutrients. Its flower emits a strong floral fragrance that is similar enough to the odor of feces to effectively entice dung beetles and carrion beetles into its pollination trap. The *Hydnora africana* flower is a temporary trap with downward facing hairs that impede escape efforts and funnel pollinators further into the plant where they inadvertently collect and/or deliver pollen. A couple days later, the flower opens enough for pollen-covered beetles to escape and potentially fall for the pollination trap yet again.

The Amazon water lily (*Victoria amazonica*) has a highly specialized pollination system involving its night-blooming flower that only lasts for 2-3 days. The flower's white color on the first night makes it easier to find in low light conditions, which is important because it can only be pollinated during the first bloom. The flower has a pineapple fragrance that attracts scarab beetles (*Cyclocephala castanea* and *Cyclocephala hardyi*) and an increased internal temperature that attracts the beetles while also assisting the aroma's dispersal. The flower keeps beetles from leaving before pollen is produced by providing a warm place to stay, starchy food bodies, and closing its petals around them until the next bloom. By the second bloom, the anthers have dusted the beetles with pollen and the flower has changed to a pinkish-red or purple color, become scentless, and stopped producing extra heat. These changes help the pollination system because beetles are not attracted to flowers that have already been pollinated.

Mucuna holtonii is a bat-pollinated plant that is highly specialized for attracting the Commissaris's long-tongued bat (*Glossophaga commissarisi*) using a structural feature that complements the high-frequency sound of the bat's echolocation. The plant's nocturnally blooming flowers also emit a scent strong enough to travel long distances and get a flying bat's attention. The distinguishing feature its flower uses to attract bats is a concave banner petal that indicates the location of available nectar by reflecting a bat's echolocation back towards the source. The specialized banner petal of each flower only becomes raised after it has finished producing pollen and the nectar reward. Commissaris's long-tongued bats must land on each flower to access nectar and the bat's light weight is enough to trigger a powerful release of pollen onto the bat's back. Each flower's banner petal alters its acoustic reflection when its shape changes after a bat visits, which the bats learn is an indication a flower does not have nectar.

The dormancy of deciduous trees during harsh conditions, such as winter in areas that have snow or ice, is an appropriate response to environmental stresses. Dormancy is an appropriate response to freezing conditions because expanding water in a tree's trunk, branches, and leaves would cause considerable damage that could have a negative impact on survivability. Another environmental stressor that is overcome through dormancy and diverting water to the roots is the difficulty of obtaining water when the ground is frozen. Deciduous trees avoid ice damage in their leaves and don't spend resources on their upkeep when photosynthesis would not be happening anyway by dropping their leaves when winter is coming. The conservation of energy for surviving wintery conditions and sprouting new leaves when conditions improve is aided by Abscisic acid preventing the division of cells. Dormancy is made possible by a tree's appropriate reaction to changing environmental conditions including the increase or decrease of temperatures and amount of daylight received.

The wood frog (*Lithobates sylvaticus* or *Rana sylvatica*) has defenses against freezing conditions that make it possible for them to survive when sixty-five percent of their body water is frozen. Their heart stops beating and they stop breathing during a state of suspended animation that continues throughout these cold periods. An accumulation of urea in the wood frog's tissues combined with large amounts of glucose transported into cells from the liver serve as protection against excessive shrinkage and ice formation. Cells have extra time for implementing defenses against freezing conditions because nucleating agents in the blood, such as special proteins, certain minerals, and bacteria, assist ice growth in extracellular water. Delicate tissues and organs are protected from the damaging effects of ice because much of the water is kept away from these areas while protections against freezing temperatures are occurring. The wood frog also has an accurate internal trigger that restarts their heart and breathing when they begin to thaw.

Monarch butterflies (*Danaus plexippus*) are well-known for their ability to undertake an incredible 3,000-mile-long migration that is made possible because of the Methuselah generation. Migrating to a warmer environment is critical to the survival of Monarchs because they can't survive the cold winters of northern climates. The Methuselah generation differs from the other three generations that happen in a given year because these Monarchs live for up to eight months instead of the typical six weeks. Environmental cues in late summer or early fall trigger the development of the Methuselah generation, which will overwinter in an appropriate location such as the oyamel fir forest in Mexico. It is critical for Monarch butterflies to have conditional instincts that match their conditional lifespans, so there is not a uniform response in regards to whether they should migrate. Purposefulness is evident with the existence of Methuselah generations because environmental factors determine the longevity of generations rather than lifespans being random or remaining static.

The various forms of mimicry show that genetic changes associated with characteristics of plants and animals can have the potentially beneficial outcome of them resembling another feature in their environment. Some instances of visual mimicry involve organisms with elaborate similarities to a particular aspect of its environment allowing it to effectively blend in with its surroundings. Batesian mimicry is a deceptive defense wherein a relatively harmless organism shares characteristics with a model organism that their predators recognize as dangerous and/or unpalatable. This deception only works against the mimic's predator if that predator recognizes the warning features associated with the dangerous organism, which means the model needs to live within proper proximity. Pouyannian mimicry is represented through plants achieving pollination when a male insect attempts copulation with a flower that is mimicking characteristics of potential female mates within the pollinator's same species. Effective mimicry utilizing characteristics of a model in the proper proximity is an intriguing similarity between plants and animals.

The satanic leaf-tailed gecko (*Uroplatus phanaticus*) is able to blend in with dead leaves in the forests of Madagascar because it gains visual mimicry through an assortment of physical characteristics. Mimicry of a dead leaf is accomplished with their distinctive flat tail combined with lines on their back and tail that resemble the midrib and secondary veins of a leaf. The variety of colors expressed in populations of satanic leaf-tailed geckos including shades of brown, orange, and yellow are representative of the colors displayed by decaying leaves in their environment. Matching the coloration of dead leaves is sometimes enhanced with the addition of black dots and/or greenish blemishes that look like moss. Some satanic leaf-tailed geckos even have notches along the edge of their tail resembling parts of a dead leaf that have rotted away or been nibbled by insects. Their mimicry is effective because they spend their days among dead leaves during the day and hunt insects at night.

The ant-mimicking treehopper (*Cyphonia Clavata*) lives in South American rainforests and other nearby locations in which their ant-like protrusion provides protection from predators. These treehoppers have a wing-like protrusion extending from their pronotum that adequately mimics an ant's body plan and the defensive spines found protruding from the propodeum of some ant species. The mimicked ant body plan looks backwards when compared to the treehopper's body plan, which gives the impression of an ant defensively backing up while the treehopper moves forward. What sets the treehopper's protrusion apart from evolutionary feats of other insects has to do with it being a body plan innovation rather than modification of an existing anatomical feature. A body plan innovation is noteworthy because it shows the emergence of a novel feature while most evolutionary examples involve the reduction or loss of functionality for an anatomical feature. This mimicry is effective because the ant-like protrusion is modeled after an aggressive tree ant that lives nearby.

The black milk snake (*Lampropeltis triangulum gaigeae*) is a nonvenomous subspecies of milk snake that mimics coral snakes and is found in the mountainous areas of Costa Rica and Panama. Juvenile black milk snakes have coloration that matches local populations of coral snakes while also mimicking the patterns of their rings. Being the largest known milk snake is problematic for maintaining effective mimicry because adult milk snakes keep growing longer until they are unlikely to be mistaken for a coral snake. A length discrepancy exists because coral snakes grow from 7-inch juveniles into 2-4-foot adults while black milk snakes grow from 12-16-inch juveniles into 4-6-foot adults. Black milk snakes avoid the problem of becoming too long for mimicking coral snakes by undergoing a color change that gradually makes them completely black as they transition into adulthood. Their color change is effective because it is completed before they surpass the length of adult coral snakes and mimicry protects them as juveniles.

The mirror orchid (*Ophrys speculum*) lives throughout the Mediterranean region and has a specialized pollination system that involves tricking male scoliid wasps (*Dasyscolia ciliate*) with multiple forms of mimicry. Mirror orchids attract male wasps because they resemble potential mates by having a similar shape, hairs, and a centered glossy patch of blue that looks like sky reflecting off wings. Mimicry of a potential mate is enhanced with a floral scent that closely matches mating pheromones used by female wasps. Emitting pheromones that match potential mates of an available pollinator helps more than visual mimicry because male wasps might become too excited from the pheromones to focus on visual inconsistencies. Pollination occurs when a male wasp attempts to copulate with multiple mirror orchids because they inadvertently collect pollen during the first attempt and pollinate the next flower that tricks them. Tricking pollinators with mimicry is an efficient use of the plant's resources since they don't produce costly rewards such as nectar.

The warty hammer orchid (*Drakaea livida*) is found growing in the sandy soils of Western Australia and is pollinated by males from a single species of thynnid wasp (*Zaspilothynnus nigripes*). Warty hammer orchids have a labellum that visually resembles a wingless female thynnid wasp and emits the correct combination of chemicals to replicate potent pheromones that will attract the males. Pollination occurs because males visit multiple orchids and trigger the hinged stem that brings them into contact with the flower's reproductive structures when they attempt to carry away the labellum. Mimicking characteristics of female wasps attracts the male wasps, but this peculiar pollination system only functions properly because it appropriately parallels the mating behaviors of thynnid wasps. The flightless female thynnid wasps climb up a blade of grass, release pheromones, and wait for a male to find them. Males emerge from the ground, track the pheromones, pick up the female wasp, and copulate while carrying her to a food source.

The South American leaf fish (*Monocirrhus polyacanthus*) is a small predatory fish that uses mimicry to stealthily approach unsuspecting prey. Mimicry of a dead leaf is made possible through an oval-shaped body, being the correct size, and having a filament protruding from their lower jaw that resembles a leaf stalk. The appropriate coloration of these fish further decreases suspicion of potential prey and they can even adjust their coloration to match the dead leaves found in that environment. Their mimicry of a dead leaf is strengthened through behaviors such as swimming slowly with its head down in a similar manner to how a dead leaf drifts through water. South American leaf fish maintain the appearance of a drifting dead leaf as they slowly approach potential prey with transparent pectoral fins that help mask their attempt to swim closer. Their aggressive mimicry is effective because they live in areas that have an accumulation of dead leaves due to minimal water flow.

The spicebush swallowtail (*Papilio troilus*) is a butterfly found in North America that benefits from various defensive traits throughout its lifecycle. Spicebush caterpillars are sometimes overlooked by predators because they have the appearance of bird droppings from the time they hatch until they are too large to effectively continue this ruse. Their next phase of defense is mimicking a snake's head, which is achieved with a shift to green coloration, fake eyespots, and sometimes lifting their head like a snake does. The color change to yellow or orange during the prepupae stage helps to camouflage these caterpillars against the leaf litter while they find a place to pupate. A spicebush swallowtail's chrysalis mimics a green or brown leaf depending on environmental factors such as whether that generation needs to enter a state of diapause during the winter months. The palatable adult phase has its own defense against predators by means of their resemblance to the unpalatable pipevine swallowtail (*Battus philenor*).

Aggressive mimicry involves predators diminishing the chance of prey detecting them by having characteristics that help them resemble a nonthreatening and/or unassuming feature in their environment. This form of mimicry allows predators to ambush prey when they come within striking distance due to there not being a strong enough signal for prey to recognize the danger. Some predators are effective at catching prey through aggressive mimicry by having specific features that appropriately mimic the prey's sustenance, which helps trick unsuspecting prey into coming within striking distance. Aggressive mimicry relies on proximity in the same way that defensive mimicry does because prey won't recognize characteristics as benign if there is not an appropriate model in the environment. This means tricking prey into approaching can require characteristics to have a high degree of specificity because the potential victims are quite familiar with features associated with sources of sustenance. Some instances of aggressive mimicry can double as defensive camouflage against their own predators.

The false cleanerfish (*Aspidontus taeniatus*) is mostly found in coral reef margins and their mimicry of bluestreak cleaner wrasse (*Labroides dimidiatus*) provides access to prey that trust the cleaning process. The bluestreak cleaner wrasse is recognized and trusted by a variety of larger fish because the wrasse's cleaning process helps remove problems such as externally attached parasites and dead skin. This trust is exploited because false cleanerfish have appropriate coloration changes throughout their stages of development that accurately match developmental coloration changes of a local population of bluestreak cleaner wrasse. Mimicry of morphological characteristics is assisted by copying the dance a bluestreak cleaner wrasse performs as an attempt to initiate the cleaning process when a fish visits the cleaning station. False cleanerfish mimic the cleaning process of the bluestreak cleaner wrasse but occasionally bite the fin of unsuspecting fish instead. The infrequency of attacks helps because fish keep trusting cleaning stations, which also means false cleanerfish maintain protection through mimicry.

The American bolas spider (*Mastophora hutchinsoni*) has adult females that prey on the smoky tetanolita (*Tetanolita mynesalis*) and bristly cutworm (*Lacinpolia renigera*). They hunt these moths using a ball of sticky silk on the end of a line of silk that is rapidly swung in a pendulum motion at approaching prey. Moths are covered in tiny brown scales that make them difficult to capture with normal spider silk because these scales easily detach, which allows the moth to escape from webs. Theses spiders have overcome this problem because their makeshift silk bolas is coated with a specialized liquid that can seep through the moth's scales and stick directly to the exoskeleton. Bolas spiders emit a mixture of chemical components that simultaneously mimics the pheromones used by both species of moth. Cutworms are typically active earlier than the smoky tetanolita and the spider's chemical mixture adjusts appropriately by decreasing the components needed for mimicking cutworm moth pheromones as the night progresses.

The orchid mantis (*Hymenopus coronatus*) uses lightning-fast reflexes to capture unsuspecting insects that mistake it for a nectar-producing flower. It is not surprising that insects are fooled because orchid mantises have flower-like characteristics such as lobes on their hind legs that are remarkably similar in appearance to flower petals. The main attractant of orchid mantises is their bright coloration because their white bodies typically have pink, yellow, or orange accents that indicate a nectar reward to a pollinator's brain. Orchid mantises are not in danger of having accent colors that are inappropriate for their current surroundings because they can slowly adjust their coloration to better match the environmental conditions. They look like a generic flower rather than a specific flower, which is beneficial because this increases the variety of pollinating insects that are likely to come within striking range. Orchid mantises are excellent at attracting victims within range because their vivid coloration is more appealing to pollinators than real flowers.

The spider-tailed horned viper (*Pseudocerastes urarachnoides*) lives in western Iran and uses refined caudal luring to get their prey within striking distance. These vipers are aptly named considering the knob-like structure at the end of their tail and elongated, outward growing scales near the tip of the tail convincingly mimic a spider. Spider-tailed horned vipers have coloration that provides excellent camouflage for their rocky surroundings with the exception of the spider-like lure. The color difference makes caudal luring possible because prey can see the appetizing spider while the rest of the viper's camouflaged body does not alert potential prey to the danger. Most of the viper's body remains motionless while the end of its tail is gently moved from side to side to help attract the attention of hungry birds. Small birds such as warblers that come close enough to peck at the spider-like lure are in for an unpleasant surprise when the viper strikes them in under a second.

CHAPTER FOUR:
Symbiotic Relationships

SYMBIOTIC RELATIONSHIPS ARE powerful demonstrations of balance occurring in an ecosystem when multiple species benefit from long-lasting interconnected bonds that sometimes involve specialized genetic changes. Facultative mutualism involves symbiotic relationships found on one end of the spectrum of interdependence wherein all members can survive without their mutualistic partners. The formation of a mutualistic relationship is challenged by competitive advantages being easier for natural selection to fix into a population than a genetic change that provides a cooperative opportunity. This is because a genetic change that improves survivability in antagonistic relationships is more likely to be routinely tested by the environment.

Natural selection can influence the spread of mutations affecting competitive or cooperative relationships, but the latter sometimes involves organisms interacting in an obscure way before benefits to survivability are realized. Cooperative interactions between multiple species are positively reinforced through increased survivability, so the formation of mutualistic relationships relies on organisms having mutually beneficial aspects while also living near each other.

Narrow-mouthed frogs (Microhylidae) are a family of frogs that are sometimes known to live together with large spiders such as tarantulas. A significant problem facing these frogs is the danger they face in their respective environments due to an abundance of animals that prey on them such as snakes and geckos. Threats to survivability are decreased when frogs are living in a tarantula's territory because these large spiders aggressively protect their nests when animals such as predators of the frog approach. Frogs living in a tarantula's nest have a steady source of sustenance since they will readily eat the ants attracted to leftover food from victims of the spider. Tarantulas benefit from the presence of narrow-mouthed frogs because consumed ants are one less threat to the spider's eggs, which is helpful since ants are major predators of these eggs. A chemical cue helps tarantulas recognize narrow-mouthed frogs as evidenced by these frogs being picked up, examined, and put down unharmed.

Sea otters (*Enhydra lutris*) can help bring balance to an entire ecosystem through their mutualistic relationship with giant kelp (*Macrocystis pyrifera*). Dense populations of kelp create forests that many aquatic species use as a nursery and a place to hide from predators. A major threat to the existence of kelp forests is the inability to protect itself against hungry sea urchins destroying entire forests due to population growth being out of control. Sea otters prevent an excessive amount of kelp from being eaten because they are natural predators of sea urchins and consequently keep the population of this herbivorous threat under control. A big contributing factor to why sea otters are so effective at controlling the population growth of sea urchins is a high metabolism that causes them to eat a lot. Having a high metabolism helps keep sea otters warm in the cold waters of their habitat while also causing the preservation of kelp forests that keep them safe.

The greater honeyguide (*Indicator indicator*) is a bird that has a mutualistic relationship with humans in sub-Saharan Africa despite a lack of training from people or the bird's parents. People wanting to locate honey can get a local honeyguide's attention with a specialized shout, whistle, or other sound that has been passed down multiple generations for this purpose. A honeyguide leads the way to a nearby beehive via a series of stops on perches visible to their human companions coupled with a call to indicate their location. The bird uses a call to indicate a nest is nearby so the people can perform their part of the arrangement by calming bees with smoke and breaking open the hive. This partnership allows people to forgo the tedious task of locating beehives that are hidden in crevices of tall trees. The greater honeyguide's potential reward includes grubs and leftover beeswax since they are one of the few birds able to digest wax.

It is not surprising that clownfish and sea anemones are associated with each other in the wild considering the benefits involved with their mutualistic relationship. Clownfish lightly touch parts of their body to an anemone to acclimate themselves and/or help make proper adjustments to the specialized mucus protecting them from the sea anemone's venomous tentacles. An anemone provides a safe place for clownfish to live and lay eggs because predators of clownfish can't easily bypass the stinging cells and might become sustenance for the anemone. The host anemone gains protection from the presence of clownfish because they eat small invertebrates and chase away butterfly fish, which are all potential threats to the sea anemone. A host sea anemone has better access to sustenance because some fish are lured by the bright colors of the clownfish and fecal matter from the clownfish also provides nutrients. Clownfish eat leftover scraps from prey the anemone has caught and dead tentacles from their host.

Some species of ants and aphids have a well-matched relationship based around ants' attraction to sugar and aphids' propensity for sucking sap out of plants and excreting a sugar-rich honeydew. Ants enjoy the sugary substance so much that they stroke aphids to stimulate the release of honeydew and gather a group of aphids for ready access to a steady supply. Ants protect their source of sugar by attacking predators of aphids such as ladybugs and accommodate the needs of aphid eggs during winter by storing them in their own nests. Aphids are carried to a new food source when winter has ended or a host plant is depleted of resources. Aphids usually develop wings to fly to new sustenance when a plant is depleted of resources or the aphid population becomes too dense. Ants stop aphids from flying away by biting wings off if they have already developed and even produce a chemical that interferes with the development of wings.

The thick-footed morel (*Morchella crassipes*) is a soil fungus that displays the various elements of farming in their mutualistic interactions with the bacteria *Pseudomonas putida*. The hyphae of thick-footed morels provide a way for their mutualistic bacteria to travel through soil more efficiently than their competitors. The mutualistic bacteria get an additional benefit from travelling through soil with the help of thick-footed morels' hyphae because carbon is transferred from the host fungus to *P. putida*. The phase in the relationship that is reminiscent of harvesting happens periodically and consists of an influx of carbon for the fungus during the process of absorbing the mutualistic bacteria. Some of the collected carbon is stored in hard-walled sclerotia, which form in some species of fungi as a reservoir of resources that helps during times of unfavorable environmental conditions. The presence of mutualistic bacteria results in sclerotia forming away from the bacteria, which contrasts with the indiscriminate growth that happens when bacteria are absent.

The Hawaiian bobtail squid (Euprymna scolopes) has a mutualistic relationship with a bioluminescent bacterium (*Aliivibrio fischeri*) involving an assortment of specialized aspects. The Hawaiian bobtail's light organ houses the bioluminescent bacteria, has functions similar to a photoreceptor, and manipulates the bacteria's produced light with proteins known as reflectins that refract incident light. This light organ allows the squid to hunt prey and avoid predators at night because adjustments made to the intensity of light from the bacteria results in camouflage known as counter-illumination. Elimination of the squid's shadow through matching the incident light is accomplished by controlling the ink sac in their mantle in a manner reminiscent of an eye's diaphragm. *A. fischeri* thrive within the light organ because the squid provides an appropriate mixture of sugar and amino acids. Most *A. fischeri* are ejected from Hawaiian bobtail squids daily and the gene expression for bioluminescence is only activated when the correct density of these bacteria is reached around dusk.

Juvenile Hawaiian bobtail squids have a system for obtaining the first batch of bioluminescent bacteria they need for counter-illumination because *A. fischeri* are not passed from parents to their offspring. Obtaining *A. fischeri* involves the secretion of a mucus from pores near the light organ in response to a component found in the cell wall of bacteria known as peptidoglycan. *A. fischeri* outcompete motility-challenged bacteria in attempts to navigate through the mucus because they have flagella to help counteract a relentless current created by ciliated cells in the light organ. Another obstacle bacterium must overcome to enter the light organ is the presence of a microbicidal enzyme that is extremely effective at killing bacteria. *A. fischeri* are unaffected by this hostile environment because it undermines the destructive effects by capturing hydrogen peroxide before it is used by the enzyme. These defensive measures used by Hawaiian bobtail squids help ensure that only their mutualistic partners make it into the light organ.

Obligate symbiotic relationships are on the opposite end of the spectrum of interdependence from facultative relationships because one or more members are reliant on their partnership for survival. The emergence of genetic changes that can result in the formation of obligate mutualistic relationships involves a level of specificity that surpasses most examples of adaptability, coevolution, and cospeciation. The formation of obligate mutualistic relationships requires an extreme level of specificity because the emergence of relevant genetic changes is unforgiving in terms of timing and proximity. A solution to the need for appropriate timing and proximity is an existent relationship between the soon-to-be obligate partners. However, an existent relationship makes the problem of specificity more acute because it further decreases the number of genetic changes that can result in a successful obligate relationship. The number of cooperative relationships that an organism maintains is typically lower than the number of possible obligate mutualistic relationships in the immediate vicinity or within a reasonable distance.

Lichen is a composite organism consisting of a combination of at least one fungus and one or more photosynthetic organisms such as green algae and/or cyanobacteria. Lichens stand apart from other mutualistic relationships because the bonding of these organisms results in structures, chemicals, and synthesized substances that the component organisms can't replicate outside of their mutualistic partnership. Fungi create the lichen's structure, which protects algae and/or cyanobacteria from the environment while also collecting water and nutrients for photosynthetic partners to fulfil their own role. The algae and/or cyanobacteria contribute by combining the collected nutrients and water with sunlight through photosynthesis to create enough sugar for itself and its fungal partner(s). This mutualistic relationship results in resilience that makes it possible for lichen to live in a wide range of environments such as the Arctic tundra, deserts, and rocky coasts. A lichen can't exist without the combination of its component organisms and some component organisms can't survive without their mutualistic partnership.

The golden jellyfish (*Mastigias papua etpisoni*) has mostly lost its tentacles and sting while living in the relatively safe Jellyfish Lake and is reliant on mutualistic single-celled dinoflagellates called zooxanthellae. Juvenile golden jellyfish obtain zooxanthellae by engulfing thousands of these algae-like organisms, which take up residence in the jellyfish's tissues and give their host a golden glow. Zooxanthellae keep their host alive because golden jellyfish get their energy from carbon products produced through their mutualistic partner's photosynthesis. Golden jellyfish have a daily routine that involves propelling themselves across the lake to mirror the sun's movement, which results in providing their zooxanthellae with sunlight they need for photosynthesis. These jellyfish further accommodate their mutualistic partner's sunlight requirements because they provide even exposure to sunlight by rotating counter-clockwise while swimming across the lake. Golden jellyfish's reliance on their photosynthetic partners is evidenced through a rise in water temperature in 1998 killing zooxanthellae and the adult jellyfish population dropping to zero in response.

Each species of entomopathogenic nematodes from the genus *Steinernema* has a mutualistic relationship with a single species of gram-negative bacteria from the genus *Xenorhabdus*. The bacteria have access to sustenance and an environment for them to multiply rapidly once their host nematode has infected an insect larva and secreted a mixture of anti-immune proteins. *Xenorhabdus* bacteria need a nematode host for survival because they can't locate and infect insects on their own and are unable to survive in the soil where the insects live. Nematodes benefit from this relationship because the bacteria produce toxins that kill the insect in a day or two and make the environment viable for the nematode's growth and reproduction. The bacteria provide sustenance for their mutualistic partner when they break down the insect's tissues and they also prevent competition for the available resources by producing antibiotic compounds. Bacteria colonize a nematode host's intestinal tract after being ingested by an infective juvenile and subsequently become quiescent.

Some siboglinid worms can survive the extreme conditions found around hydrothermal vents thanks to an obligate mutualistic relationship with internal bacteria. Siboglinid worm adults' reliance on sustenance provided by mutualistic bacteria is evident because these worms completely lack components of a digestive system such as a mouth, digestive tract, and anus. The bacteria enter siboglinid worms through the worm's skin and reside in a special storage organ known as a trophosome. Siboglinid worms provide mutualistic bacteria with chemicals needed for chemosynthesis due to a special type of hemoglobin that makes it possible to transport oxygen and sulfide at the same time. Symbiotic bacteria use sulfide as an energy source for carbon dioxide fixation through the reductive tricarboxyclic (rTCA) cycle and/or Calvin cycle to create organic molecules usable by the siboglinid worm. The ability to use the Calvin cycle, rTCA cycle, or both at the same time allows bacteria to continue producing organic molecules regardless of how much oxygen is available.

Pea aphids (*Acyrthosiphon pisum*) have an obligate mutualistic relationship with internal bacteria (*Buchnera aphidicola*) that likely started 160 to 280 million years ago. The aphids' nutritionally poor diet of plant sap does not provide the essential amino acid arginine, which means aphids must obtain it from another source. Their symbiotic bacteria remedy this deficiency in the aphid's diet by using the amino acid glutamine to create arginine through biosynthesis and sharing the essential amino acid with their host. Excess glutamine found in phloem is transported to *B. aphidicola,* which is then used in the process of synthesizing essential and nonessential amino acids that are shared with the aphid. Bacteriocytes are specialized cells with membranal molecules that are well-suited for transporting relevant amino acids between mutualistic bacteria housed within and the host. Female aphids pass bacteria-containing bacteriocytes to their offspring through the reproductive process, which is critical since the aphid's life cycle usually involves reproduction without the involvement of a male.

Some species of lower termites have an obligate mutualistic relationship with protozoa living in their gut. This partnership dates back at least 100 million years because a termite with protozoa spilling out of its damaged abdomen was encased in amber dated to around this age. Protozoa benefit from their relationship with a host termite because they have a safe environment to live in and gain access to cellulose since termites ingest wood regularly. These partners need each other for survival because termites can't digest cellulose that they can access, and protozoa can't freely access cellulose for which they have enzymes to break down. Protozoa produce enzymes known as cellulases to effectively break the extremely durable sugar molecules cellulose is composed of down to sugars the termite and protozoa can digest. Termite larvae and termites that have lost their symbionts due to recently molting get new obligate mutualistic partners into their gut by ingesting protozoa that are found in termite feces.

Acacia trees in Central America and acacia ants (*Pseudomyrmex ferrugineus*) have a symbiotic relationship in which the trees cause the ants' participation in the partnership to become obligate. Acacia ants adamantly protect their host tree by attacking herbivores and nearby vegetation that would compete with their host for nutrients and/or sunlight. A major benefit for the ants is not needing to search for food because sustenance is provided by their host tree in the form of extrafloral nectar and food bodies. Acacia ants can only get carbohydrates from *Acacia* trees' extrafloral nectar because sucrose molecules are broken into smaller sugars that ants can digest by means of a plant-produced enzyme called invertase. Acacia ants could make their own invertase if extrafloral nectar did not also contain an enzyme called chitinase, which blocks the ants' ability to produce their own invertase enzyme. Blocking invertase production through a change to the ants' phenotype stops Acacia ants from getting sustenance from another source.

Some species of rainforest plants in the genus *Piper* have a mutualistic relationship with a species of ants (*Pheidole bicornis*). *Piper* plants benefit from their symbiotic relationship with ants because these ants protect their host plants by attacking encroaching vines and herbivorous insects, and unintentionally helping to destroy fungal pathogens. The ants are provided with a nest because the plants' petioles have enclosed chambers that ants can easily access. Ants expand these naturally formed nests for additional space to live by removing the pith tissue inside a plant's stems. These mutualistic ants seldomly have to forage on the ground for food because their host plant produces food bodies inside the petiole chambers that are appropriate for the ants' needs. *Piper* plants don't waste resources on the formation of food bodies because production is triggered when the mutualistic ants are nearby. Host plants also get a small portion of their nitrogen intake by absorbing organic matter from the ants' nesting areas.

An unusual mutualistic relationship exists in northwestern Borneo, Indonesia between the fanged pitcher-plant (*Nepenthes bicalcarata*) and the diving ant (*Camponotus schmitzi*). Diving ants are only associated with the fanged pitcher-plant and the host accommodates their mutualistic partners with hollow tendrils. A diving ant can safely swim and submerge themselves because *N. bicalcarata* actively maintains its pitcher fluid at lower acidity levels. Diving ants protect their host from pitcher-destroying weevils and prevent the loss of some nutrients by attacking the insects that are trying to escape the pitcher plant's fluid. An even more important service the ants provide to their host is the removal of large prey that would rot if they remained in the pitcher's fluid. Host plants also benefit from their ant symbionts in the form of nutrient-rich droppings. The fanged pitcher-plant gets a significant increase in their nitrogen intake with the presence of mutualistic ants, which results in the plant having more leaves than those without ant symbionts.

Some species of leafcutter ants have a mutualistic relationship with a species of fungus that dates back millions of years. The leafcutter worker ant's ability to carry twenty times its body weight allows it to carry relatively large sections of leaf to the nest after clipping it from the tree. Leafcutter ants are unable to directly eat foraged leaves because tropical leaves have chemical toxins that are usually effective at deterring herbivores from doing extensive damage to a plant's foliage. Chemical toxins in the leaves are ineffective at deterring worker ants from clipping leaves from trees and there are instances of entire trees being defoliated in less than twenty-four hours. Leafcutter ants gather leaves they can't directly eat because the leaves are broken down and used as a substrate for a mutualistic fungus. The fungus secretes an enzyme that decomposes the gathered leaves and subsequently provides the ants' source of sustenance in the form of nutritious structures known as gongylidia.

Minims are the smallest ants in a leafcutter colony and have the pivotal role of breaking down leaves to create a substrate for the mutualistic fungus. Fungal farms face a recurring threat in the form of an aggressive fungus known as Escovopsis that would quickly destroy fungus farms if effective antibiotics were not available. Some species of leafcutter ants have a type of worker that has a special gland for secreting phenylacetic acid, which kills fungal infections without harming the farmed fungus. The other species of leafcutter ants obtain effective antibiotics from beneficial filamentous bacteria that are housed in modified areas of worker ants' exoskeletons. Ant-fungus mutualism is obligate because the fungus provides the ants with sustenance and the fungus needs help with propagation because it can't sexually reproduce. Species of leafcutter ants reliant on bacteria for production of effective antibiotics are in a three-way obligate mutualistic relationship because each organism is reliant on the other two for survival.

Bucket orchids belong to the genus *Coryanthes* and each species has a pollination system that is dependent on mutualistic bees from the tribe Euglossini. Male orchid bees are attracted to fragrant oils produced by bucket orchids because the bees collect these oils to create a chemical blend that is instrumental in attracting a mate. Successful pollination of a bucket orchid starts with a mutualistic bee losing its grip while collecting the flower's fragrant oils and subsequently falling into the flower's bucket of plant-produced liquid. The only means of escape for the wet bee is an appropriately sized tunnel containing the flower's reproductive system. The orchid glues pollen sacs to the bee's back and the tightened walls of the tunnel hold the bee while the glue dries, which can take up to 45 minutes. The pollination process is completed when the mutualistic bee falls into another orchid wherein a hook attached to the top of the tunnel removes the pollen sacs.

Male orchid bees collect fragrant oils from mutualistic orchids that they combine with compounds from other sources such as fungi and leaves to create an impressive and attractive chemical blend. A male orchid bee uses specialized tarsal brushes on their modified forelegs to scrape the sides of the orchid's labellum so it can collect the fragrant oils produced there. They transfer the fragrant compounds to pouches on their back legs, which is the reason why some lose their grip on the flower and fall into the bucket of liquid. Female orchid bees are not attracted to bucket orchids even though their male counterparts use fragrances from these flowers in their mating ritual. This is because fragrant compounds gathered from bucket orchids constitute a small portion of the chemical blend used to attract and impress a potential mate. The use of chemical mixtures for attracting mates allows male bees to differentiate themselves from competitors and demonstrate their suitability to female bees.

Mycorrhizal fungi have the distinction of taking part in mutualistic relationships involving the root systems of 90% of land plants and all orchids. Mycorrhizal fungi benefit from their relationship with their host orchid because the flower's expanding root system increases the symbiont's access to organic matter that can subsequently be turned into nutrients. Increased access to these resources enables the fungal partner to reproduce faster and the symbiont shares nutrients such as carbon with their host and protects the plant from pathogens. This mutualistic partnership is obligate for all orchids because their seeds are incapable of germinating and growing to the point of leaf production without nutrients provided by their mutualistic fungi. Mycorrhizal fungi that are viable for the successful germination of each species of orchid can range in specificity from a single species to a more generalized list. The downy rattlesnake plantain (*Goodyera pubescens*) even switches between species of mutualistic fungi when environmental changes occur such as a drought.

Species of ambrosia beetles from the weevil subfamilies Scolytinae and Platypodinae have mutualistic relationships with fungi that have resulted in multiple instances of coevolution involving beetles and their farmed fungi. Fungal farms are commonly associated with the higher levels of ethanol found in weakened and dead trees because many threats to the fungi have their growth inhibited in this environment. Ethanol causes these fungal farms to grow faster which consequently decreases the wait time for initiating the reproductive process of beetles since mutualistic fungi are their only source of sustenance. Ambrosia beetles help their fungal partners by transporting them to new sources of sustenance, which these species of fungi can't do themselves given that no free-living examples have been found. The beetles accomplish this transportation by collecting spores from the fungal farm and storing them in specialized structures called mycangia. The high level of diversity found in mycangia of farming beetles happened because this mutualistic partnership convergently evolved multiple times.

The glassy-winged sharpshooter (*Homalodisca vitripennis*) is a large leafhopper insect with two bacterial symbionts known as *Candidatus Sulcia muelleri* and *Candidatus Baumannia cicadellinicola*. This three-way obligate mutualistic partnership makes it possible for glassy-winged sharpshooters to subsist on a nutritionally poor diet of xylem sap taken from a variety of plants. These sharpshooters have specialized host organs called bacteriomes to house their symbionts. Glassy-winged sharpshooters have two different bacteriomes that are easily distinguishable by color. The red bacteriome only houses Baumannia while the yellow one is occupied by both symbionts. *S. muelleri* contributes to the symbiotic triangle by producing a majority of the essential amino acids that are limited in the host's diet of xylem sap. *B. cicadellinicola* fulfills a different role in the symbiotic relationship depending on which bacteriome it is located in. These significant functions include the synthesis of cofactors and vitamins for their host and co-symbiont. *B. cicadellinicola* also synthesizes essential amino acids not produced by *S. muelleri*.

The citrus mealybug (Planococcus citri) relies on endosymbiotic bacteria known as *Candidatus Tremblaya princeps* and *Candidatus Moranella endobia* to produce nutrients that are deficient in the phloem they eat. The bacteria living in the bacteriocytes of the mealybug's bacteriome are special because *M. endobia* is an endosymbiont located within the cytoplasm of *T. princeps*. *T. princeps* contributes to the survival of the host mealybug and its own symbiont because it handles part of the synthesis of essential amino acids. *M. endobia* also helps with the synthesis of essential amino acids, but the bacteria can't encode complete pathways for these amino acids even when they work together. The successful synthesis of nutrients is possible despite this conundrum because there are 22 bacterial-like genes within the mealybug's genome that were not transferred from either symbiont. Needing three organisms to synthesize essential nutrients shows a potential outcome of convergent genome reductions that are common with endosymbionts such as *M. endobia* and *T. princeps*.

Many species of yucca moths within the

genera *Tegeticula* and *Parategeticula* have interdependent mutualistic relationships with various species of *Yucca* plants. The larvae of these yucca moths are dependent on the sustenance provided by their mutualistic partner in the form of seeds. Yucca moths survive winter by burrowing underground and creating individual cocoons that they emerge from in time for the blooming of yucca plants. Female yucca moths deliberately pollinate yucca plants with a specialized mouth appendage that is well-suited for collecting and transferring pollen for their mutualistic partner. Yucca plants provide no immediate sustenance rewards to entice their pollinators, which is not a problem for their pollinators because yucca moths only live for about three days. If a yucca moth inspects a yucca flower and detects the scent of another female, then the number of eggs she lays is reduced or she searches for another flower. This behavior makes sense because yucca plants abort flowers with too many eggs.

Fig wasps in the family Agaonidae are well-known for their two-way obligate relationship that has formed with the fig plants they pollinate. Hundreds of these highly specialized mutualistic relationships have emerged thanks to coevolution and cospeciation events that started over sixty million years ago. The beginning of a female fig wasp's lifecycle involves mating, deliberately or inadvertently collecting pollen, and exiting through holes that are formed by males born in the same fig. The next stage consists of finding a fig to lay their eggs in while carrying the pollen in specialized pouches or wherever the pollen became attached while the female escaped. They access the inside of a fig by squeezing through a tiny hole known as an ostiole, which can cause their wings and antennae to break off or become damaged. Once inside, each mutualists' reproductive cycle has the proper means to continue when the female fig wasp finishes her task of laying eggs and pollinating the fig.

There are around 850 species of fig plants that each have a small number of specialized fig wasp species suitable as mutualistic partners. Pollination of the fig's inward-blooming flowers is unlikely to occur without a mutualistic fig wasp since the flowers lack a reward for most potential pollinators and are difficult to access. Providing mutualistic fig wasps with their only source of sustenance is enough of a reward to ensure the perpetual cycle of the fig's reproductive process. Fig plants get the attention of their mutualistic fig wasps with a chemical signal that indicates they are ready for pollination. This chemical signal is for the females since male wasps spend their entire lives in a single fig. Males are wingless and their only roles are to mate and create tunnels for the females to escape through. The death of wasps inside the fig benefits the plant because it has an enzyme called ficin that turns the dead wasps into nutrients.

CHAPTER FIVE:
Purposeful Precursors

DETERMINING WHETHER MUTATIONS emerged through undirected randomness or purposefulness is irrelevant if reproductive life was unable to exist in the first place. The precursors that made the emergence of life possible having some degree of purposefulness does not automatically mean purposefulness is involved in the evolutionary process. However, it makes sense that purposefulness would persist beyond the emergence of life and might play a role in some aspects of the evolutionary process. It is important to note the existence of aspects in our universe that are strongly associated with purposefulness does not invalidate the existence of completely random aspects and vice versa. The examples herein are not meant to affirm any worldviews, but rather are intended to point out the involvement of purposefulness in our universe that made life possible. Furthermore, the existence of purposeful aspects that are difficult to explain in a universe based on pure randomness rather than also including some purposefulness does not indicate where purposefulness originated.

One way in which purposefulness could've been involved in making the emergence of life possible is through the constants of the universe that have the appearance of being finely tuned. Purposefulness might be involved because some universal constants have outcomes with extremely rigid tolerance levels in regards to conditions being suitable for the emergence of life. One possible explanation for universal constants having the exact outcomes seen in our universe is there were no alternative outcomes for the outcomes of these constants. Removal of other possible outcomes invalidates the apparent precision of universal constants and removes the involvement of purposefulness. However, it is unlikely universal constants were guaranteed to support life without the involvement of purposefulness because a universe based on pure randomness would lack a tendency toward any outcome. The cumulative coincidences become absurd when considering the likelihood of every universal constant having a guaranteed outcome that supports the emergence of life without a degree of purposefulness being involved.

A finely-tuned aspect of the universe that played a critical role in the establishment of conditions that made the eventual emergence of life possible is the cosmological constant. The cosmological constant represents dark energy, which accounts for approximately 68% of the universe's mass-energy and also has an anti-gravitational effect that causes the expansion of the universe to accelerate. The cosmological constant being finely-tuned to one part in 10^{120} is part of the reason that we live in a flat universe, which means the universe's expansion won't reverse. Life would not have been possible if dark energy caused the early universe to expand too quickly because galaxies and stars couldn't form under those conditions. Conversely, life couldn't emerge if dark energy caused the early universe to expand too slowly because the universe would have collapsed back in on itself. The universe expanded at the correct rate to allow the emergence of life despite the leeway for deviation being 1 in 10^{55}.

Another finely-tuned aspect of our universe that made the emergence of life possible involves how much weaker the gravitational force is in comparison to the electromagnetic force. The ratio of the electromagnetic force to the gravitational force is 4.15×10^{42} and the max deviation for this ratio to remain in the proper range is one part in 10^{40}. If the electromagnetic force's strength changed in relation to the gravitational force's strength enough to exceed the threshold of fine-tuning, then the chance of any biological life emerging would plummet. A ratio between these fundamental forces that exceeds the life-permitting range would significantly decrease the amount of carbon, oxygen, and other heavier elements formed through nuclear fusion within stars. The formation of heavier elements within stars relies on conditions such as having a sufficiently high internal temperature needed for the nuclear fusion of elements needed for life. Also, insufficient temperatures would've stopped supernovae from making heavier elements available for the formation of planets.

The formation of the first proteins is another topic in which there are astronomically low odds of success if undirected randomness was the only causal factor involved. This would require the first proteins to form through a random sequence of amino acids because protein synthesis uses ribosomal proteins in the process of building proteins for each cell. Getting an appropriate protein through a random combination of amino acids is problematic given that the successful formation of a protein requires a specific sequence of amino acids. There are twenty amino acids that can attach at each site in the sequence, so the probability of success decreases exponentially as the length of the required sequence increases. A sequence made from only ten amino acids has over ten trillion possible combinations and this is not representative since the average protein for Archaea is 283 amino acids. It is ridiculously unlikely that a protein formed through pure randomness was appropriate for the first cell.

The formation of functional proteins through random combinations had further challenges including amino acids having left-handed and right-handed forms that each have about an equal chance of occurring. This is a problem for the probability of proteins forming without purposefulness because proteins can't function if there is a mixture of left-handed and right-handed amino acids in the sequence. There is a one in forty chance of the correct amino acid being at each site in a sequence if the selection between using left-handed and right-handed variants was random. Purposefulness might've alleviated this problem through a recognition system that favored one form over the other so that right-handed amino acids weren't used in the formation of the earliest proteins. Another problem facing the formation of the earliest proteins is the non-elimination of sequences that failed to produce a functional protein. This means the odds of success for the earliest proteins couldn't change over time with subsequent failures to form functional proteins.

The microscopic building blocks needed for the formation of the first reproductive cell existing in the proper timeframe while also having the correct proximity is highly unlikely with pure randomness. Achieving the appropriate proximity is complicated by obstacles in the environment. However, proximity is less problematic for pure randomness despite the size difference of a planet and biological building blocks because relocation is easier to accomplish than preserving successful combinatorial outcomes. The various building blocks of life would most likely be destroyed by environmental factors or through degeneration over time before any compatible building blocks appeared in the same location. The successful formation of a protein through randomness is irrelevant if this protein does not persist until the other building blocks needed for a cell are in the appropriate proximity. This problem is exasperated when considering that every protein needed for the various aspects of a functioning cell needed appropriate proximity and timing to result in a functional, biological cell.

A plausible explanation that invalidates all of the probabilistic problems facing the emergence of life is the idea of multiple universes existing rather than just the one we live in. It is critical for these parallel universes to have randomly determined values for physical constants that can vary widely from the finely tuned outcomes found in our universe. Many parallel universes are likely devoid of biological life because their physical constants are not within the appropriate range or other astronomical odds were not overcome. However, the parameters needed for life-permitting conditions happened at least once since abiogenesis and evolution occurred in our universe. Essentially, the existence of a universe with conditions suitable for the formation of habitable planets and subsequently reproductive life would be statistically inevitable given enough parallel universes. If there are enough parallel universes to make the involvement of probability inconsequential for the emergence of life, then purposefulness is irrelevant for explaining how these astronomical odds were overcome.

The existence of an infinite number of parallel universes would not remove the need for purposefulness in our universe because probabilities don't affect the innate functionality of biological building blocks. Overcoming the odds against a universe with life-permitting conditions is futile if the functionalities needed for reproductive life are unaccounted for. The formation of functionalities for each of the biological building blocks needed for an abundance of biodiversity might involve mechanisms that create functionalities specifically useful for the emergence of life. The existence of this type of mechanism contradicts the idea of a universe based on purely undirected randomness because random forces can realistically only affect functionalities that already exist. If the specific functionalities needed for various biological processes always existed, then this points to a degree of inevitability that does not logically mesh with a universe lacking any purposefulness. However, it is understandable that these biological functionalities are usually taken for granted because life eventually emerged in our universe.

The building blocks needed for the emergence of life have a combination of functionality and emergent properties, which means they have characteristics that are not associated with their individual components. Proteins, lipids, and nucleic acids each have functionalities that complement each other in the formation of replicating cells associated with life as we know it. Functionality existing at each level of complexity is a strong indicator of purposefulness because emergent properties related to biological systems typically have characteristics that are useful for a specific purpose. Multiple layers of functionality present in reproductive life include amines and carboxylic acids combining into amino acids that subsequently connect to form proteins, which are functional in their own right. Another important aspect of proteins is the spontaneous folding of polypeptide chains into a three-dimensional shape when the correct sequence of amino acids is achieved. Spontaneously folding into a shape is critical because this shape determines the functions a protein can perform in biological systems.

The astronomically low odds of a protein forming through random combinations of amino acids and the existence of emergent properties are irrelevant if valid sequences didn't exist. This means each protein's shape and functionality within a biological system needed to be associated with a specific sequence of amino acids before proteins could operate in the first cell. A wide variety of functionalities for plant and animal proteins are connected to valid sequences and more probably exist considering how many species have gone extinct in Earth's history. An abundance of orphan genes has been found for species that haven't gone extinct and it stands to reason that extinct species would've had their own orphan genes. Purposefulness has likely coexisted with undirected randomness at least since the formation of valid protein sequences. The existence of valid sequences along with appropriate amino acids that can combine together to bring these proteins into fruition is one of the strongest pieces of evidence for purposefulness.

The linking together of amino acids into a functional protein is like linking together alphabetic letters to create a meaningful word. There are many possible combinations of alphabetic letters with limited results that make sense in written communication just like the limited number of sequences that can result in functional proteins. Words needing agreed upon meanings for reliable written communication is similar to proteins needing purposeful functionality before their formation fulfills a purpose within cells. A specific sequence of letters needing particular meanings attributed to it is similar to specific sequences of amino acids needing protein functionality attached to them. Agreed upon meanings of words did not arise through undirected randomness in a similar way to the advent of functionality for proteins needing more input than undirected randomness can accomplish.

However, proteins are much more complicated than alphabetic words and intended meanings can feasibly be attached to any combination of letters while the sequence of a valid protein is static.

A major hurdle for the emergence of reproductive life in a universe without a degree of purposefulness is the formation of a standardized process when a beneficial biological process emerges. A lack of purposefulness would necessitate use of random interactions of building blocks to attain the correct order of events needed to accomplish a task such as protein synthesis. This functionality must occur in a specific order at specific locations within the cell which requires more input into the process than the mere availability of appropriate building blocks. Repetition of a beneficial process would likely require recognition of when the correct set of steps have been accomplished so this process happens consistently in the future. Therefore, a transition from random interactions to consistent steps indicates a degree of purposefulness. This is where the question arises of how much recognition is possible in a system based on undirected randomness because biological processes need consistent steps of action alongside appropriate building blocks.

The standardization of biological processes was necessary at an early stage of reproductive life since mitosis, phagocytosis, and protein synthesis are critical to the onset of life and biodiversity. Mitosis requires the completion of a strict sequence of complex stages known as prophase, metaphase, anaphase, and telophase after the cell has finished the steps of interphase. A cell also needs a standardized process such as phagocytosis for uptake and utilization of nutrients available in the environment. Protein synthesis is also needed to become standardized at some point because attaining proteins through random combinations of amino acids would be too inefficient for reproductive life. Protein synthesis needed some method of transcription to copy instructions for constructing the correct protein from an available nucleic acid. Accomplishing this first step was inconsequential until there was a method of building protein by translating the instructions copied from the nucleic acid. The synthesized protein then needed to go where it is needed within the cell.

There needs to be awareness at the microscopic level for cells to respond to stimuli and recognize particular needs including which proteins currently need synthesizing. Protein synthesis is an excellent indicator of a cell's level of awareness considering an appropriate ratio of proteins is maintained despite the immense number and variety of proteins involved. Components of a cell having the ability to detect and respond to stimuli makes phenomena such as phenotypic plasticity possible through an appropriate response to temperature and other external cues. Another example of awareness involves the multiple enzymes responsible for fixing damage done to DNA through mechanisms known as base excision repair and nucleotide excision repair. Repair mechanisms help show why purposefulness is a likely explanation for some evolutionary changes because the cell is aware that the new mutation is self-induced rather than from random damage. Awareness also makes it possible for each protein to detect which components it is meant to interact with or disregard.

Awareness is another emergent property that is better explained through the existence of a degree of purposefulness rather than a universe only based on undirected randomness. Purposefulness is evident with some aspects of awareness such as the exact instructions on the DNA strand consistently being copied for protein synthesis despite the many other sections available. An experiment that shows the effectiveness of awareness and is attributable to purposeful changes involves the E. coli strain FC40. When lactose is plentiful in their environment, about 20% of the cells from this strain gain the ability to utilize lactose when they couldn't previously. The duality of undirected randomness and purposefulness in our universe is still seen in regard to awareness even though it is usually difficult to determine the source of positive outcomes. An example of malfunctioning awareness that could be characteristic of many more instances of awareness in the absence of purposefulness is demonstrated by the abnormal immune responses of autoimmune diseases.

A biological process encapsulating the importance of consistent and reliable awareness is recognition of pathogens when compared to the organism's own healthy tissue through the body's innate and adaptive immune systems. The immune system responds to the recognition of foreign substances known as antigens through the use of large proteins known as antibodies, which detect and neutralize these antigens. The immune system also has another means for identifying and eliminating pathogens through the use of various white blood cells such as mast cells, natural killer cells, and phagocytic cells. Innate immune responses involve non-specific defenses that can help quickly against a wide range of internal and external threats. The innate immune system sometimes lacks a response for a pathogen and triggers antigen presentation in which the adaptive immune system develops a receptor that is effective against foreign pathogens. Immunological memory is stored information for previously encountered pathogens that protects against future infections and is the reason why vaccines are useful.

The necessity for awareness at the microscopic level is mirrored by modern technological systems such as desktop computers and gaming consoles. Functionality of a desktop computer is derived from internal components including the processor, motherboard, desktop memory, graphics card, hard drive, and usually a case to protect all of these parts. However, a sound system and/or visual display of some sort such as a monitor are essential for accomplishing tasks on these systems despite the system's full functionality being at hand. Actions could still occur on these systems with a mouse and/or keyboard with decreased efficiency, but progress would represent random results rather than intentional outcomes. This inefficiency would be amplified for proteins and other building blocks lacking an appropriate level of awareness because they lack a means of compensating for this deficiency. If antibodies lacked the ability to detect nearby antigens, then they are unlikely to fulfill their functionality of eliminating the threat because the probability of interaction decreases.

The hypothetical mechanism I propose that ties the information presented in this book into a coherent set of evidence involves some concepts borrowed from quantum mechanics. Quantum biology is not a unique or controversial idea since it has already been established that concepts from quantum mechanics play a role in some biological processes. However, some instances that involve quantum mechanics are not always agreed upon since a process could be an example of quantum biology or another explanation might work instead. Quantum biology is a young field of study so the processes that have already been explored and purposeful evolution will have to wait an indeterminable length of time for verification. It makes sense that quantum mechanics is involved in biological systems since all biological components can be reduced down to particles that are governed by quantum mechanics. Therefore, it is unsurprising that aspects of quantum mechanics such as quantum entanglement continue affecting some functional biological systems to some degree.

A photon is a unit of electromagnetic radiation such as visible light that has wave-particle duality and acts as a means to transport energy. Photons sometimes demonstrate properties of waves as seen by their ability to make interference patterns, which would not occur if photons only had the characteristics of particles. On the other hand, photons sometimes demonstrate properties of particles as evidenced through the photoelectric effect, which involves its ability to dislodge electrons when the correct threshold frequency is reached. This duality made life possible because energy carries energy from the sun that has a high enough frequency to cause the photoelectric effect that makes photosynthesis possible.

It is important that photons with a lower frequency have less energy because higher levels of energy are capable of damaging organisms. People protect themselves from ultraviolet light with sunscreen and use lead to block x-rays and gamma rays because the high frequency of these waves become harmful with too much exposure.

A photon's ability to carry information that can be decoded is an established fact that is taken advantage of everyday by millions of electronic devices. Electromagnetic waves are so useful for quickly transmitting information that people have been using them in a wide range of electronic devices such as radios, televisions, and cellphones. Short wavelengths associated with high frequencies such as those of gamma rays are capable of bypassing many materials because the waves are too small to interact with the obstacle. The wavelengths of radio waves have a similar lack of interaction with a number of obstacles due to being too large and slow to interact with these materials. The light given off by the sun that we can perceive is typically reflected or absorbed by materials because it has an appropriate wavelength for interaction. This means photons with extremely low frequencies could carry information while passing through obstacles until they interact with a system with a compatible receiver.

A foundational idea that helps explain what is occurring in some biological processes is the simultaneous coexistence of all possible outcomes for a system's physical state while it is unknown. A wave function represents the probability distribution of possible outcomes for a system's definite physical state that are contained within its superposition. A system needs a definite physical state for interaction with another system so the initiation of interaction results in a wave function collapse that reveals the system's definite physical state. The reality of particles being in superposition is demonstrated with the double slit experiment, in which individual particles pass through both slits. A pattern emerges that indicates the probabilistic outcomes of the wave function and individual particles can interfere with themselves after passing through both slits, which results in an interference pattern. However, the very act of detecting the particle to see this phenomenon is enough to collapse the wave function because this is a form of interaction.

Quantum entanglement is a physical

phenomenon that is reliant on the existence of superpositions and involves pairs of particles that are irreversibly linked. Interactions that affect one of the entangled particles such as a measurement of one of its physical properties results in the collapse of this particle's superposition. This interaction with the particle also directly affects the result of the wave function collapse of its entangled partner with a guaranteed inverse result. This correlation occurs between entangled particles regardless of how far apart the entangled particles are when the wave function collapses for either particle. This indicates that an entangled pair of particles are parts of the same superposition, which makes sense because either their wave function has collapsed or it hasn't. The entanglement of particles can be represented by a math problem involving two numbers that have a sum of zero. Needing a positive number and its equivalent negative number means that both values are always determined simultaneously.

European robins (*Erithacus rubecula*) have been studied due to their accuracy during migration wherein they can fly from Scandinavia to the exact areas in the Mediterranean that their ancestors used.

Specialized funnels, hoods, strong magnets, wavelengths of light, ink, and scratch sensitive paper were used in various tests to help figure out what is affecting their uncanny sense of direction. The tests indicated the migratory orientation of European robins is impacted by the Earth's magnetic field and the frequency associated with blue light. The process that helps these birds navigate involves the cryptochrome protein in their right eye producing magnetically sensitive radical pairs when it interacts with blue light. These radical pairs are electrons that produce chemical reactions based on tiny variations in the Earth's magnetic field with resultant navigationally relevant signals sent to the brain. Quantum entanglement is proposed to be involved with the radical pair electrons in this process since entangled particles are quite sensitive to magnetic fields.

A widely accepted example of quantum biology involves the ridiculously high energy transfer efficiency rate found in regard to the conversion of light into the chemical energy that occurs during photosynthesis. The first step of this conversion happens when solar energy is transferred to an electron within the chlorophyll of a photosynthetic organism. This energy is then transported to a reaction center so that this energy is stored for the organism to use later.

It seems straightforward and lacks any need for reference to quantum mechanics until the multiple paths available to the energy-carrying electron are taken into consideration. The random pathing expected for these electrons in a system based purely on classical mechanics is inadequate for explaining the high energy transfer efficiency rate observed during this transference. Random pathing is ruled out because these electrons take the most efficient route to the reaction center by remaining in superposition until all paths are sampled and the best option is found.

Biophotons are photons with a range of frequencies associated with ultraviolet and visible spectrums that are constantly emitted by most living cells, which includes non-luminous microorganisms such as bacteria. Biophotons were an unknown phenomenon until Alexander Gurwitsch discovered them in 1923 because their intensity is 1,000 times less than a human's naked eye can perceive. DNA is the storage location for biophotons, and it is a significant source of biophotons produced by organisms as evidenced by the cessation of biophoton emission when the nucleus is removed. Biophotons have laser-like coherence with a coherence time that is far superior to lasers and this high degree of order makes it possible for the transmission of information. The highly-structured and organized frequency of biophotons explains how they can carry information found on the DNA strand with clarity for extended periods of time. These information-laden biophotons are able to quickly spread along light-sensitive microtubules made from tubulin proteins and even travel outside the organism.

The idea of information carried by biophotons being used for rapid intracellular and intercellular communication has already been verified through observations of plants, bacteria, and kidney cells. The information carried by biophotons can affect organisms in a meaningful way because cells act as senders and receivers of these particles. The regulation of 100,000 biochemical reactions per second in each cell is likely helped by biophotons because classical mechanics are insufficient by themselves to explain the timing and sequential order of these reactions. There are even observations corroborating the communication of biophotons between cells that are mechanically, chemically, and electrically isolated from each other. One such example involves human cells synchronizing internal chemical processes while separated in lab conditions. Separated populations of E. coli have stimulated the growth of each other when conditions allow populations to interact via biophotons. A similar result was observed with plants wherein stimulation from a growing plant increased the cell division of nearby plants by 30%.

Stress having a direct impact on organisms and the ability to communicate this stress to nearby biological systems leads us to the proposed mechanism underlying purposeful evolution. The mechanism involves organisms adapting to environmental conditions via appropriate genetic changes through combining awareness of relevant factors and using biophotons to send and receive information about stress-inducing circumstances. This means preexisting mutations are not attributable to purposeful evolution when they only become advantageous after environmental conditions change because meaningful causality is lost. Purposeful evolution is possible because information being carried by biophotons represents a universal language given that it is derived from data stored on a DNA strand. This means the information transmitted via biophotons is compatible with any lifeforms that have biological systems governed by a similar nucleic acid. A universal language is a critical component because the existence of communication that is only compatible with members of the same species would undermine purposeful evolution's ability to influence obligate symbiotic relationships.

Furthermore, an indication that cells are aware of stressful environmental conditions is the noticeable increase in biophoton emission when comparing a cell that is stressed to a healthy cell. The opposite has also been observed wherein reduced free radicals in cells have a direct impact through a decrease in biophoton emission. An injured leaf has demonstrated this difference since the cells at the location of an injury release more biophotons than other cells of the plant. Another example has been found concerning poisoned algae having an increased biophoton output when compared to their healthy counterparts. Humans also show a marked increase in biophoton emission under various stresses including exposure to ultraviolet radiation and carcinogens in cigarette smoke. Changes in biophoton emission might have functional relevance such as being a distress signal. An increase in biophoton emission could help purposeful evolution because this increase makes it more likely that information concerning a stressful factor is communicated to organisms in the vicinity.

This mechanism explains how purposeful evolution could have contributed to some of the transitions from unicellular life to multicellular life. This might seem presumptuous to say considering these transitions occurred millions of years ago. However, this form of evolution has been observed in a lab setting when conditions were appropriate for the transition and it demonstrated a distinct lack of gradualism. The results of the experiment involved populations of the unicellular green alga *Chlamydomonas reinhardtii* convergently evolving in the presence of *Paramecium tetraurelia,* which is a unicellular predator. It took 50 weeks for two out of five populations to evolve the ability to become multicellular in the presence of a predator. The populations of algae without a predator did not become multicellular during this timeframe, which is mostly a nonfactor given the small sample size. These rapid genetic changes were appropriate because transitioning from being unicellular to multicellular resulted in the unicellular predator being too small to eat the algae.

An apparent conundrum for purposeful evolution is the lack of purposeful changes given that populations could hypothetically mutate every generation based on purposeful changes of other organisms in their environment. The likely reason this whirlwind of evolutionary changes doesn't occur is a threshold of change that must be met before organisms experience purposeful mutations. This threshold seems to increase in direct proportion to the complexity of the species that is experiencing stressful conditions, which explains why bacteria and insects can evolve quickly. The DNA must also experience appropriate mutations for purposeful evolution and the extent of randomness involved with this process is debatable. There is also arguably a range for this threshold wherein some members of a population need less stress to reach this threshold while others require more stress. This unfortunately makes it difficult to get consistent results when studying purposeful evolution because a population might go extinct before both reach their threshold for change and undertake appropriate mutations.

These thresholds have arguably been demonstrated through lab experiments studying the rate at which E. coli strain FC40 reverts to a phenotype that makes the exploitation of lactose possible. These studies illuminate FC40's tendency to revert to this phenotype when lactose is the only source of carbon. It was thought that reversion would happen when strains of E. coli were experiencing nutrient starvation regardless of whether lactose was available. This is unlikely given that the rate of reversion does not change when cells are experiencing nutrient starvation while also not having access to lactose. Jeffrey Stumpf, Anthony Poteete, and Patricia Foster found that true reversions can arise in a single step within the first five days of experiencing nutrient starvation while lactose is present. Their findings also showed a sharp rise in probability for reversion on the fifth day. This is expected for purposeful evolution because the threshold for change must first be reached and this threshold encompasses some variation.

In conclusion, the balance that has emerged in many ecosystems is likely attributable to a nonrandom factor such as purposeful evolution. This balance is noticeable through amazing adaptations that have occurred, evolutionary arms races, and cospeciation events that make obligate mutualistic relationships possible. The validity of this form of evolution would explain evolution's inconsistency including the lack of gradualism when conditions change. For example, the rate of mutations should not change just because a population is separated from their peers and faces new challenges, but this is exactly what has happened repeatedly. This distinct lack of gradualism is mirrored by the periods of stasis and periodic abrupt changes associated with punctuated equilibrium and the appropriate speciation events that occur after massive extinctions. Undirected randomness also lacks explanatory power when it comes to the intricacy of various forms of defensive and offensive mimicry. It is my hope that the information and examples throughout this book have met the burden of proof.

Made in the USA
Monee, IL
05 November 2026

67375676R00090